Undergraduate Texts in Mathematics

Undergraduate Texts in Mathematics

Apostol: Introduction to Analytic Number Theory.
1976. xii, 338 pages. 24 illus.

Armstrong: Basic Topology.
1983. xii, 260 pages. 132 illus.

Bak/Newman: Complex Analysis.
1982. x, 224 pages. 69 illus.

Banchoff/Wermer: Linear Algebra Through Geometry.
1983. x, 257 pages. 81 illus.

Childs: A Concrete Introduction to Higher Algebra.
1979. xiv, 338 pages. 8 illus.

Chung: Elementary Probability Theory with Stochastic Processes.
1975. xvi, 325 pages. 36 illus.

Croom: Basic Concepts of Algebraic Topology.
1978. x, 177 pages. 46 illus.

Curtis: Linear Algebra: An Introductory Approach (4th edition)
1984. x, 337 pages. 37 illus.

Dixmier: General Topology.
1984. x, 140 pages. 13 illus.

Ebbinghaus/Flum/Thomas Mathematical Logic.
1984. xii, 216 pages. 1 illus.

Fischer: Intermediate Real Analysis.
1983. xiv, 770 pages. 100 illus.

Fleming: Functions of Several Variables. Second edition.
1977. xi, 411 pages. 96 illus.

Foulds: Optimization Techniques: An Introduction.
1981. xii, 502 pages. 72 illus.

Franklin: Methods of Mathematical Economics. Linear and Nonlinear Programming. Fixed-Point Theorems.
1980. x, 297 pages. 38 illus.

Halmos: Finite-Dimensional Vector Spaces. Second edition.
1974. viii, 200 pages.

Halmos: Naive Set Theory.
1974, vii, 104 pages.

Iooss/Joseph: Elementary Stability and Bifurcation Theory.
1980. xv, 286 pages. 47 illus.

Jänich: Topology
1984. ix, 180 pages (approx.). 180 illus.

Kemeny/Snell: Finite Markov Chains.
1976. ix, 224 pages. 11 illus.

Lang: Undergraduate Analysis
1983. xiii, 545 pages. 52 illus.

Lax/Burstein/Lax: Calculus with Applications and Computing, Volume 1. Corrected Second Printing.
1984. xi, 513 pages. 170 illus.

LeCuyer: College Mathematics with A Programming Language.
1978. xii, 420 pages. 144 illus.

Macki/Strauss: Introduction to Optimal Control Theory.
1981. xiii, 168 pages. 68 illus.

Malitz: Introduction to Mathematical Logic: Set Theory - Computable Functions - Model Theory.
1979. xii, 198 pages. 2 illus.

continued after Index

Jacques Dixmier

General Topology

Springer-Verlag
New York Berlin Heidelberg Tokyo

Jacques Dixmier
Université de Paris VI
Mathématiques UER 48
9 Quai Saint Bernard
75 Paris
France

Translated from the French by
Sterling K. Berberian
Department of Mathematics
The University of Texas
Austin, TX 78712
U.S.A.

AMS Subject Classification: 54-01

Library of Congress Cataloging in Publication Data
Dixmier, Jacques.
General topology.
(Undergraduate texts in mathematics)
Translation of: Topologie générale.
Includes indexes.
1. Topology. I. Title. II. Series.
QA611.D48613 1984 514′.322 83-20444

With 13 Illustrations

Title of the original French edition: *Topologie Générale*, © Presses Universitaires de France, Paris, 1981.

Printed in the United States of America.

9 8 7 6 5 4 3 2 1

ISBN 978-1-4419-2823-8

Contents

Introduction

This book is a course in general topology, intended for students in the first year of the second cycle (in other words, students in their third university year). The course was taught during the first semester of the 1979–80 academic year (three hours a week of lecture, four hours a week of guided work).

Topology is the study of the notions of limit and continuity and thus is, in principle, very ancient. However, we shall limit ourselves to the origins of the theory since the nineteenth century. One of the sources of topology is the effort to clarify the theory of real-valued functions of a real variable: uniform continuity, uniform convergence, equicontinuity, Bolzano–Weierstrass theorem (this work is historically inseparable from the attempts to *define* with precision what the real numbers are). Cauchy was one of the pioneers in this direction, but the errors that slip into his work prove how hard it was to isolate the right concepts. Cantor came along a bit later; his researches into trigonometric series led him to study in detail sets of points of **R** (whence the concepts of open set and closed set in **R**, which in his work are intermingled with much subtler concepts).

The foregoing alone does not justify the very general framework in which this course is set. The fact is that the concepts mentioned above have shown themselves to be useful for objects other than the real numbers. First of all, since the nineteenth century, for points of $\mathbf{R}^n$. Next, especially in the twentieth century, in a good many other sets: the set of lines in a plane, the set of linear transformations in a real vector space, the group of rotations, the Lorentz group, etc. Then in 'infinite-dimensional' sets: the set of all continuous functions, the set of all vector fields, etc.

Topology divides into 'general topology' (of which this course exposes the rudiments) and 'algebraic topology', which is based on general topology but makes use of a lot of algebra. We cite some theorems whose most natural proofs appeal to algebraic topology:

(1) let B be a closed ball in $\mathbf{R}^n$, f a continuous mapping of B into B; then f has a fixed point;
(2) for every $x \in \mathbf{S}_2$ (the 2-dimensional sphere) let $\vec{V}(x)$ be a vector tangent to $\mathbf{S}_2$ at x; suppose that $\vec{V}(x)$ depends continuously on x; then there exists an $x_0 \in \mathbf{S}_2$ such that $\vec{V}(x_0) = 0$;
(3) let U and V be homeomorphic subsets of $\mathbf{R}^n$; if U is open in $\mathbf{R}^n$, then V is open in $\mathbf{R}^n$.

These theorems cannot be obtained by the methods of this course, but, having seen their statements, some readers will perhaps want to learn something about algebraic topology.

The sign ► in the margin pertains to theorems that are especially deep or especially useful. The choice of these statements entails a large measure of arbitrariness: there obviously exist many little remarks, very easy and constantly used, that are not graced by the sign ►.

The sign * signals a passage that is at the limits of 'the program' (by which I mean what has been more or less traditional to teach at this level for some years).

Quite a few of the statements have already been encountered in the First cycle. For clarity and coherence of the text, it seemed preferable to take them up again in detail.

The English edition differs from the French by various minor improvements and by the addition of a section on normal spaces (Chapter 7, Section 6).

CHAPTER I

Topological Spaces

After reviewing in §1 certain concepts already known concerning metric spaces, we introduce topological spaces in §2, then the simplest concepts associated with them. For example, one has an intuitive notion of what is a boundary point of a set E (a point that is 'at the edge' of E), a point adherent to E (a point that belongs either to E or to its edge), and an interior point of E (a point that belongs to E but is not on the edge). The precise definitions and the corresponding theorems occupy §§4 and 5. Separated topological spaces are introduced in §6; on first reading, the student can suppose in what follows that all of the spaces considered are separated.

1.1. Open Sets and Closed Sets in a Metric Space

1.1.1. Let E be a set. Recall that a *metric* (or 'distance function') on E is a function d, defined on E × E, with real values ≥ 0, satisfying the following conditions:

(i) $d(x, y) = 0 \Leftrightarrow x = y$;
(ii) $d(x, y) = d(y, x)$ for all x, y in E;
(iii) $d(x, z) \leq d(x, y) + d(y, z)$ for all x, y, z in E ('triangle inequality').

(On occasion we shall admit the value $+\infty$ for a metric; this changes almost nothing in what follows.)

A set equipped with a metric is called a *metric space*. One knows that the preceding axioms imply

(iv) $|d(x, z) - d(x, y)| \leq d(y, z)$ for all x, y, z in E.

There is an obvious notion of isomorphism between metric spaces.

Let E′ be a subset of E. Take the restriction to $E' \times E'$ of the given distance function d on $E \times E$. Then E′ becomes a metric space, called a *metric subspace* of E.

1.1.2. Examples. The ordinary plane and ordinary space, with the usual Euclidean distance, are metric spaces. For $x = (x_1, x_2, \ldots, x_n) \in \mathbf{R}^n$ and $y = (y_1, y_2, \ldots, y_n) \in \mathbf{R}^n$, set

$$d(x, y) = ((x_1 - y_1)^2 + \cdots + (x_n - y_n)^2)^{1/2}.$$

One knows that d is a metric on $\mathbf{R}^n$, and in this way $\mathbf{R}^n$ becomes a metric space, as do all of its subsets. In particular $\mathbf{R}$, equipped with the metric $(x, y) \mapsto |x - y|$, is a metric space.

1.1.3. Definition. Let E be a metric space (thus equipped with a metric d), A a subset of E. One says that A is *open* if, for each $x_0 \in A$, there exists an $\varepsilon > 0$ such that every point x of E satisfying $d(x_0, x) < \varepsilon$ belongs to A.

1.1.4. Example. Let E be a metric space, $a \in E$, ρ a number ≥ 0, A the set of all $x \in E$ such that $d(a, x) < \rho$. Then A is open. For, let $x_0 \in A$. Then $d(a, x_0) < \rho$. Set $\varepsilon = \rho - d(a, x_0) > 0$. If $x \in E$ is such that $d(x_0, x) < \varepsilon$, then

$$d(a, x) \leq d(a, x_0) + d(x_0, x) < d(a, x_0) + \varepsilon = \rho,$$

thus $x \in A$.

The set A is called the *open ball with center a and radius* ρ. If $\rho > 0$ then $a \in A$; if $\rho = 0$ then $A = \varnothing$. In the ordinary plane, one says 'disc' rather than 'ball'.

1.1.5. In particular, let a, b be real numbers such that $a \leq b$. The interval (a, b) is nothing more than the open ball in $\mathbf{R}$ with center $\frac{1}{2}(a + b)$ and radius $\frac{1}{2}(b - a)$. One verifies easily that the intervals $(-\infty, a)$, $(a, +\infty)$ are open. This justifies the expression 'open interval' employed in elementary courses.

1.1.6. Theorem. *Let* E *be a metric space.*

(i) *The subsets* $\varnothing$ *and* E *of* E *are open.*
(ii) *Every union of open subsets of* E *is open.*
(iii) *Every finite intersection of open subsets of* E *is open.*

The assertion (i) is immediate.

Let $(A_i)_{i \in I}$ be a family of open subsets of E, and let

$$A = \bigcup_{i \in I} A_i, \qquad A' = \bigcap_{i \in I} A_i.$$

Let us show that A is open. Let $x_0 \in A$. There exists $i \in I$ such that $x_0 \in A_i$. Then there exists $\varepsilon > 0$ such that the open ball B with center x_0 and radius

ε is contained in A_i. *A fortiori*, $A \supset B$. Thus A is open. Assuming I is finite, let us show that A′ is open. Let $x_1 \in A'$. For every $i \in I$, there exists $\varepsilon_i > 0$ such that the open ball with center x_1 and radius ε_i is contained in A_i. Let ε be the smallest of the ε_i. Then $\varepsilon > 0$, and the open ball with center x_1 and radius ε is contained in each A_i, hence in A′. Thus A′ is open.

1.1.7. Let us maintain the preceding notations. If I is infinite, $\bigcap_{i \in I} A_i$ is not always open. For example, in **R**, the intersection of the open intervals $(-1/n, 1/n)$ for $n = 1, 2, 3, \ldots$ reduces to $\{0\}$, thus is not open.

1.1.8. Definition. Let E be a metric space, B a subset of E. One says that B is *closed* if the subset $E - B$ is open.

1.1.9. Example. Let $a \in E$, $\rho \geq 0$, B the set of $x \in E$ such that $d(a, x) \leq \rho$. Then B is closed. For, let $x_0 \in E - B$. Then $d(a, x_0) > \rho$. Set $\varepsilon = d(a, x_0) - \rho > 0$. If $x \in E$ is such that $d(x_0, x) < \varepsilon$, then

$$d(a, x) \geq d(a, x_0) - d(x_0, x) > d(a, x_0) - \varepsilon = \rho,$$

therefore $x \in E - B$. Thus $E - B$ is open, consequently B is closed.

The set B is called the *closed ball with center a and radius ρ*. One has $a \in B$. If $\rho = 0$ then $B = \{a\}$. In the ordinary plane, one says 'disc' rather than 'ball'.

1.1.10. In particular, let a, b be real numbers such that $a \leq b$. The interval $[a, b]$ is nothing more than the closed ball in **R** with center $\frac{1}{2}(a + b)$ and radius $\frac{1}{2}(b - a)$. One verifies easily that the intervals $[a, +\infty)$ and $(-\infty, a]$ are closed. This justifies the expression 'closed interval' employed in elementary courses. One also sees that an interval of the form $[a, b)$ or $(a, b]$, with $a < b$, is neither open nor closed.

1.1.11. Theorem. *Let* E *be a metric space.*

(i) *The subsets* $\varnothing$ *and* E *of* E *are closed.*
(ii) *Every intersection of closed subsets of* E *is closed.*
(iii) *Every finite union of closed subsets of* E *is closed.*

This follows from 1.1.6 by passage to complements.

1.1.12. Example. Let E be a metric space, $a \in E$, $\rho \geq 0$, S the set of all $x \in E$ such that $d(a, x) = \rho$. Then S is closed. For, let A (resp. B) be the open ball (resp. closed ball) with center a and radius ρ. Then $E - A$ is closed. Since $S = B \cap (E - A)$, S is closed by 1.1.11(ii).

The set S is called the *sphere with center a and radius ρ*. If $\rho = 0$ then $S = \{a\}$.

In **R**, a sphere of radius > 0 reduces to a set with 2 points. In the ordinary plane, one says 'circle' rather than 'sphere',

1.1.13. Let E be a metric space. On comparing 1.1.6 and 1.1.11, one sees that E (and similarly $\varnothing$) is a subset that is both open and closed. This is exceptional: in the most common examples of metric spaces, it is rare that a subset is both open and closed (cf. Chapter X).

On the other hand, although it is easy to exhibit examples of subsets that are either open or closed, it should be understood that a subset of E chosen 'at random' is in general neither open nor closed. For example, the subset **Q** of **R** is neither open nor closed.

1.1.14. Theorem. *Let* E *be a set, d and d' metrics on* E. *Suppose there exist constants $c, c' > 0$ such that*

$$c\,d(x, y) \le d'(x, y) \le c'\,d(x, y)$$

for all $x, y \in$ E. The open subsets of E *are the same for d and d'.*

Let A be a subset of E that is open for d. Let $x_0 \in$ A. There exists an $\varepsilon > 0$ such that every point x of E satisfying $d(x_0, x) < \varepsilon$ belongs to A. If $x \in$ E satisfies $d'(x_0, x) < c\varepsilon$, then $d(x_0, x) < \varepsilon$, therefore $x \in$ A. This proves that A is open for d'. Finally, one can interchange the roles of d and d' in the foregoing.

1.1.15. However (with the preceding notations) the balls and spheres of E are in general different for d and d'. For example, for $x = (x_1, \ldots, x_n) \in \mathbf{R}^n$ or $\mathbf{C}^n$, and $y = (y_1, \ldots, y_n) \in \mathbf{R}^n$ or $\mathbf{C}^n$, set

$$d(x, y) = (|x_1 - y_1|^2 + \cdots + |x_n - y_n|^2)^{1/2},$$
$$d'(x, y) = |x_1 - y_1| + \cdots + |x_n - y_n|,$$
$$d''(x, y) = \sup(|x_1 - y_1|, \ldots, |x_n - y_n|).$$

One knows that d, d', d'' satisfy the conditions of 1.1.14, hence define the same open subsets of $\mathbf{R}^n$. However, for d'' the open ball with center $(x_1, \ldots, x_n)$ and radius ρ is an 'open slab with center $(x_1, \ldots, x_n)$':

$$(x_1 - \rho, x_1 + \rho) \times (x_2 - \rho, x_2 + \rho) \times \cdots \times (x_n - \rho, x_n + \rho).$$

1.2. Topological Spaces

1.2.1. Definition. One calls *topological space* a set E equipped with a family $\mathcal{O}$ of subsets of E (called the *open sets of* E) satisfying the following conditions:

(i) the subsets $\varnothing$ and E of E are open;
(ii) every union of open subsets of E is open;
(iii) every finite intersection of open subsets of E is open.

One also says that $\mathcal{O}$ defines a *topology* on E.

1.2.2. For example, a metric space is automatically a topological space, thanks to 1.1.3 and 1.1.6; this structure of topological space does not change if one replaces the metric d of E by a metric d' related to d by the condition of 1.1.14.

In particular, every subset of the ordinary plane, or of ordinary space, or of $\mathbf{R}^n$, is a topological space. For a large part of the course, these are the only interesting examples we shall have at our disposal; but they already exhibit a host of phenomena.

1.2.3. Let E be a set. In general there is more than one way of choosing in E a family $\mathcal{O}$ of subsets satisfying the conditions 1.2.1. In other words, a set E may be equipped with more than one structure of topological space. For example, if one takes for $\mathcal{O}$ the family of all subsets of E, the conditions 1.2.1 are satisfied, thus E becomes a topological space called a *discrete space* (one also says that the topology of E is discrete). For another example, if one takes for $\mathcal{O}$ the family consisting only of $\varnothing$ and E, the conditions 1.2.1 are satisfied, thus E becomes a topological space called a *coarse space* (or 'indiscrete space'); one also says that E carries the *coarsest topology* (or 'indiscrete topology'). If $\mathcal{T}_1$ and $\mathcal{T}_2$ are topologies on E, $\mathcal{T}_1$ is said to be *finer* than $\mathcal{T}_2$ (and $\mathcal{T}_2$ *coarser* than $\mathcal{T}_1$) if every open set for $\mathcal{T}_2$ is open for $\mathcal{T}_1$; this is an order relation among topologies. Every topology on E is finer than the coarsest topology, and coarser than the discrete topology.

1.2.4. For example on $\mathbf{R}^n$ one can consider, in addition to the topology defined in 1.2.2, the discrete topology and the coarsest topology (and, to be sure, many other topologies). However, it is the topology defined in 1.2.2 that is by far the most interesting. Whenever we speak of $\mathbf{R}^n$ as a topological space without further specification, it is always the topology defined in 1.2.2 that is understood.

1.2.5. Definition. Let E be a topological space, A a subset of E. One says that A is *closed* if the subset E − A is open.

1.2.6. Theorem. *Let* E *be a topological space.*

(i) *The subsets* $\varnothing$ *and* E *of* E *are closed.*
(ii) *Every intersection of closed subsets of* E *is closed.*
(iii) *Every finite union of closed subsets of* E *is closed.*

This follows from 1.2.1 by passage to complements.

1.3. Neighborhoods

1.3.1. Definition. Let E be a topological space and let $x \in E$. A subset V of E is called a *neighborhood of x in* E if there exists an open subset U of E such that $x \in U \subset V$.

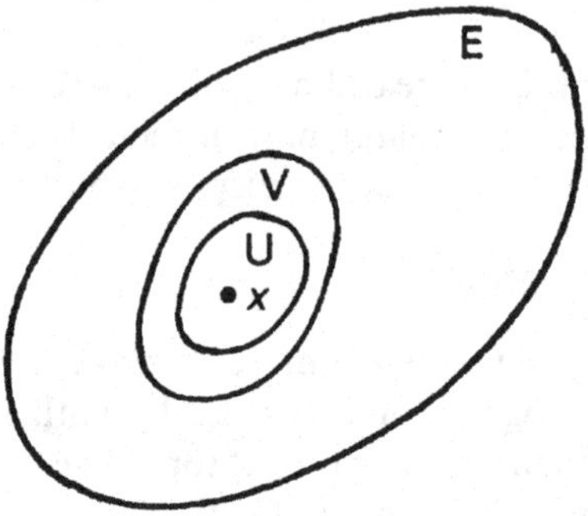

According to this definition, an *open neighborhood of* x is nothing more than an open subset of E that contains x.

1.3.2. Example. Let E be a metric space, $x \in E$, $V \subset E$. The following conditions are equivalent:

(i) V is a neighborhood of x;
(ii) there exists an open ball with center x and radius > 0 that is contained in V.

(ii) $\Rightarrow$ (i). This is clear since the ball considered in (ii) is open and contains x.
(i) $\Rightarrow$ (ii). If V is a neighborhood of x, there exists an open subset U of E such that $x \in U \subset V$. By 1.1.3, there exists $\varepsilon > 0$ such that the open ball with center x and radius ε is contained in U, thus *a fortiori* contained in V.

1.3.3. Example. In **R**, consider the subset $A = [0, 1]$. Let $x \in \mathbf{R}$. If $0 < x < 1$ then A is a neighborhood of x. If $x \geq 1$ or $x \leq 0$, A is not a neighborhood of x.

1.3.4. Theorem. *Let* E *be a topological space, and let* $x \in E$.

(i) *If* V *and* V′ *are neighborhoods of* x, *then* $V \cap V'$ *is a neighborhood of* x.
(ii) *If* V *is a neighborhood of* x, *and* W *is a subset of* E *containing* V, *then* W *is a neighborhood of* x.

Let V, V′ be neighborhoods of x. There exist open subsets U, U′ of E such that $x \in U \subset V$, $x \in U' \subset V'$. Then

$$x \in U \cap U' \subset V \cap V',$$

and $U \cap U'$ is open by 1.2.1(iii), therefore $V \cap V'$ is a neighborhood of x.
The assertion (ii) is obvious.

▶ **1.3.5. Theorem.** *Let* E *be a topological space,* A *a subset of* E. *The following conditions are equivalent*:

(i) A *is open*;
(ii) A *is a neighborhood of each of its points.*

(i) ⇒ (ii). Suppose A is open. Let $x \in A$. Then $x \in A \subset A$, thus A is a neighborhood of x.

(ii) ⇒ (i). Suppose condition (ii) is satisfied. For every $x \in A$, there exists an open subset B_x of E such that $x \in B_x \subset A$. Let $A' = \bigcup_{x \in A} B_x$. Then A′ is open by 1.2.1(ii), $A' \subset A$ since $B_x \subset A$ for all $x \in A$, and $A' \supset A$ since each point x of A belongs to B_x, hence to A′. Thus A is open.

1.3.6. Definition. Let E be a topological space, and let $x \in E$. One calls *fundamental system of neighborhoods of* x any family $(V_i)_{i \in I}$ of neighborhoods of x, such that every neighborhood of x contains one of the V_i.

1.3.7. Examples. (a) Suppose E is a metric space, $x \in E$. For $n = 1, 2, 3, \ldots$, let B_n be the open ball with center x and radius $1/n$. Then the sequence $(B_1, B_2, \ldots)$ is a fundamental system of neighborhoods of x. For, if V is a neighborhood of x, there exists an $\varepsilon > 0$ such that V contains the open ball B with center x and radius ε (1.3.2). Let n be a positive integer such that $1/n \leq \varepsilon$. Then $B_n \subset B \subset V$.

(b) Let us keep the same notations, and let B'_n be the closed ball with center x and radius $1/n$. Then $(B'_1, B'_2, \ldots)$ is a fundamental system of neighborhoods of x. For, $B_n \subset B'_n \subset B_{n-1}$, thus our assertion follows from (a).

(c) Let E be a topological space, and let $x \in E$. The set of all neighborhoods of x is a fundamental system of neighborhoods of x. The set of all open neighborhoods of x is a fundamental system of neighborhoods of x (cf. 1.3.1).

1.3.8. Let E be a topological space, and let $x \in E$. If one knows a fundamental system $(V_i)_{i \in I}$ of neighborhoods of x, then one knows all the neighborhoods of x. For, let $V \subset E$; in order that V be a neighborhood of x, it is necessary and sufficient that V contain one of the V_i (this follows from 1.3.4(ii) and 1.3.6).

1.4. Interior, Exterior, Boundary

1.4.1. Definition. Let E be a topological space, $A \subset E$, and $x \in E$. One says that x is *interior to* A if A is a neighborhood of x in E, in other words if there exists an open subset of E contained in A and containing x. The set of all points interior to A is called the *interior of* A and is often denoted $\mathring{A}$.

1.4.2. If x is interior to A, then of course $x \in A$. But the converse is not true. For example, if $E = \mathbf{R}$ and $A = [0, 1]$, then $\mathring{A} = (0, 1)$; the points 0 and 1 belong to A but are not interior to A. If $E = \mathbf{R}$ and $A = \mathbf{Z}$, then $\mathring{A} = \varnothing$.

1.4.3. Theorem. *Let* E *be a topological space,* A *a subset of* E. *Then* $\mathring{A}$ *is the largest open set contained in* A.

Let U be an open subset of E contained in A. If $x \in$ U then A is a neighborhood of x, therefore $x \in \mathring{A}$. Thus $U \subset \mathring{A}$.

It is clear that $\mathring{A} \subset A$. Let us show that $\mathring{A}$ is open. By 1.3.5, it suffices to prove that if $x \in \mathring{A}$, then $\mathring{A}$ is a neighborhood of x. Now, there exists an open subset B of E such that $x \in B \subset A$. Then $B \subset \mathring{A}$ by the first part of the proof, thus $\mathring{A}$ is a neighborhood of x.

1.4.4. Theorem. *Let* E *be a topological space,* A *a subset of* E. *The following conditions are equivalent*:

(i) A *is open*;
(ii) $A = \mathring{A}$.

(i) ⇒ (ii). If A is open, then the largest open set contained in A is A, therefore $A = \mathring{A}$ by 1.4.3.

(ii) ⇒ (i). If $A = \mathring{A}$, then A is open because $\mathring{A}$ is open (1.4.3).

1.4.5. Theorem. *Let* E *be a topological space,* A *and* B *subsets of* E. *Then* $(A \cap B)^\circ = \mathring{A} \cap \mathring{B}$.

The set $(A \cap B)^\circ$ is open (1.4.3) and is contained in $A \cap B$, hence *a fortiori* in A. Consequently $(A \cap B)^\circ \subset A^\circ$ by 1.4.3. Similarly, $(A \cap B)^\circ \subset B^\circ$, therefore $(A \cap B)^\circ \subset A^\circ \cap B^\circ$.

One has $\mathring{A} \subset A$, $\mathring{B} \subset B$, therefore $\mathring{A} \cap \mathring{B} \subset A \cap B$. But $\mathring{A} \cap \mathring{B}$ is open (1.2.1(iii)), therefore $\mathring{A} \cap \mathring{B} \subset (A \cap B)^\circ$ by 1.4.3.

1.4.6. However, in general $(A \cup B)^\circ \neq \mathring{A} \cup \mathring{B}$. For example, take $E = \mathbf{R}$, $A = [0, 1]$, $B = [1, 2]$. Then

$$A \cup B = [0, 2], \qquad \mathring{A} = (0, 1), \qquad \mathring{B} = (1, 2),$$

$$(A \cup B)^\circ = (0, 2) \neq (0, 1) \cup (1, 2).$$

1.4.7. Definition. Let E be a topological space, A a subset of E, x a point of E. One says that x is *exterior to* A if it is interior to $E - A$, in other words if there exists an open subset of E disjoint from A and containing x. The set of all exterior points of A is called the *exterior of* A; it is the interior of $E - A$. Interchanging A and $E - A$, we see that the exterior of $E - A$ is the interior of A.

1.4.8. Let E be a topological space, $A \subset E$, A_1 the interior of A, A_2 the exterior of A. The sets A_1 and A_2 are disjoint. Let $A_3 = E - (A_1 \cup A_2)$. Then A_1, A_2, A_3 form a partition of E. One says that A_3 is the *boundary* of A. It is a *closed* set, since $A_1 \cup A_2$ is open. If one interchanges A and $E - A$, then A_1

and A_2 are interchanged, therefore A_3 is unchanged: *a set and its complement have the same boundary.*

1.4.9. Example. Let A be the subset $[0, 1)$ of $\mathbf{R}$. The interior of A is $(0, 1)$, the exterior of A is $(-\infty, 0) \cup (1, +\infty)$, therefore the boundary of A is $\{0\} \cup \{1\}$.

1.5. Closure

1.5.1. Definition. Let E be a topological space, $A \subset E$ and $x \in E$. One says that x is *adherent to* A if every neighborhood of x in E intersects A. The set of all points adherent to A is called the *closure* (or 'adherence') of A, and is denoted $\bar{A}$.

1.5.2. If $x \in A$ then of course x is adherent to A; but the converse is not true. For example, if $E = \mathbf{R}$ and $A = (0, 1)$, then $\bar{A} = [0, 1]$.

1.5.3. Theorem. *Let* E *be a topological space,* A *a subset of* E. *Then* $\bar{A}$ *is the complement of the exterior of* A.

Let $x \in E$. One has the following equivalences:

$$
\begin{aligned}
x \notin \bar{A} &\Leftrightarrow \text{there exists a neighborhood of } x \text{ disjoint from A} \\
&\Leftrightarrow \text{there exists a neighborhood of } x \text{ contained in } E - A \\
&\Leftrightarrow x \text{ is interior to } E - A \\
&\Leftrightarrow x \text{ belongs to the exterior of A,}
\end{aligned}
$$

whence the theorem.

1.5.4. Theorem. *Let* E *be a topological space,* $A \subset E$, $B \subset E$.

(i) $\bar{A}$ *is the smallest closed subset of* E *containing* A;
(ii) A *is closed* $\Leftrightarrow A = \bar{A}$;
(iii) $(A \cup B)^{-} = \bar{A} \cup \bar{B}$.

In view of 1.5.3, this follows from 1.4.3, 1.4.4, 1.4.5 by passage to complements. For example, let us prove (i) in detail. The exterior of A is $(E - A)^{\circ}$ (1.4.7), that is to say, the largest open set contained in $E - A$ (1.4.3). Therefore its complement $\bar{A}$ (1.5.3) is closed and contains A. If F is a closed subset of E containing A, then $E - F$ is open and is contained in $E - A$, therefore $E - F \subset (E - A)^{\circ} = E - \bar{A}$, thus $F \supset \bar{A}$.

1.5.5. Taking up again the notations of 1.4.8, Theorem 1.5.3 shows that $\bar{A} = A_1 \cup A_3$, $(E - A)^{-} = A_2 \cup A_3$. Therefore the boundary A_3 is the intersection of the closures $\bar{A}$ and $(E - A)^{-}$.

1.5.6. Theorem. *Let* E *be a topological space,* A *a subset of* E. *The following conditions are equivalent*:

(i) *every nonempty open subset of* E *intersects* A;
(ii) *the exterior of* A *is empty*;
(iii) *the closure of* A *is all of* E.

Condition (i) means that the only open set contained in E − A is ∅, thus, by 1.4.3, that the interior of E − A is empty. This proves (i) ⇔ (ii). The equivalence (ii) ⇔ (iii) follows from 1.5.3.

1.5.7. Definition. A subset A of E satisfying the conditions 1.5.6 is said to be *dense in* E.

1.5.8. Example. **Q** is dense in **R**: for, every nonempty open subset of **R** contains a nonempty open interval, hence contains rational numbers. The complement **R** − **Q** of **Q** in **R** is also dense in **R**, because every nonempty open interval contains irrational numbers.

1.5.9. Theorem. *Let* A *be a nonempty subset of* **R** *that is bounded above, x its supremum. Then x is the largest element of* $\bar{A}$.

Let V be a neighborhood of x in **R**. There exists an open subset U of **R** such that $x \in U \subset V$. Then there exists $\varepsilon > 0$ such that $(x - \varepsilon, x + \varepsilon) \subset U$. By the definition of supremum (least upper bound), there exists $y \in A$ such that $x - \varepsilon < y \leq x$. Then $y \in U \subset V$, therefore $V \cap A \neq \varnothing$. Thus x is adherent to A.

Let $x' \in \bar{A}$. If $x' > x$, set $\varepsilon = x' - x > 0$. Then $(x' - \varepsilon, x' + \varepsilon)$ is a neighborhood of x', therefore intersects A. Let $y \in A \cap (x' - \varepsilon, x' + \varepsilon)$. Since $y > x' - \varepsilon = x$, x is not an upper bound for A, which is absurd. Thus $x' \leq x$. This proves that x is the *largest* element of $\bar{A}$.

1.6. Separated Topological Spaces

1.6.1. Definition. A topological space E is said to be *separated* (or to be a Hausdorff space) if any two distinct points of E admit disjoint neighborhoods.

1.6.2. Examples. (a) A metric space E is separated. For, let $x, y \in E$ with $x \neq y$. Set $\varepsilon = d(x, y) > 0$. Then the open balls V, W with centers x, y and radius $\varepsilon/2$ are disjoint (because if $z \in V \cap W$ then $d(x, z) < \varepsilon/2$ and $d(y, z) < \varepsilon/2$, therefore $d(x, y) < \varepsilon$, which is absurd).

(b) A discrete topological space E is separated. For, if $x, y \in E$ and $x \neq y$, then $\{x\}$ and $\{y\}$ are disjoint open neighborhoods of x, y.

(c) A coarse topological space E is not separated (if it contains more than one point). For, let $x, y \in E$ with $x \neq y$. Let V, W be neighborhoods of x, y. Then V contains an open subset U of E containing x, whence $U = E$ and *a fortiori* $V = E$. Similarly $W = E$. Thus $V \cap W \neq \varnothing$.

1.6.3. Theorem. *Let* E *be a separated topological space,* $x \in E$. *Then* $\{x\}$ *is closed.*

Let $y \in E - \{x\}$. Then $y \neq x$, therefore there exist neighborhoods V, W of x, y that are disjoint. In particular, $W \subset E - \{x\}$, therefore $E - \{x\}$ is a neighborhood of y. Thus $E - \{x\}$ is a neighborhood of each of its points, consequently is open (1.3.5). Therefore $\{x\}$ is closed.

CHAPTER II

Limits. Continuity

As mentioned in the introduction, the limit concept is one of those at the origin of topology. The student already knows several aspects of this concept: limit of a sequence of points in a metric space, limit of a function at a point, etc. To avoid a proliferation of statements later on, we present in §2 a framework (limit along a 'filter base') that encompasses all of the useful aspects of limits. It doesn't hurt to understand this general definition, but it is much more important to be familiar with a host of special cases.

The definition of limit of course brings with it that of continuity of functions: see §§3 and 4. Two topological spaces are said to be homeomorphic (§5) if there exists a bijection of one onto the other which, along with the inverse mapping, is continuous; two such spaces have the same topological properties, and one could almost consider them to be the *same* topological space. (For example, a circle and a square are homeomorphic; a circle and a line are not homeomorphic; and, what is perhaps more surprising, a line and a circle with a point omitted are homeomorphic.) A reasonable goal of topology would be to classify all topological spaces up to homeomorphism, but this seems to be out of reach at the present time.

One knows very well that a sequence does not always have a limit. As a substitute for limit, we introduce in §6 the concept of adherence value.

2.1. Filters

2.1.1. Definition. Let X be a set. A *filter* on X is a set $\mathscr{F}$ of nonempty subsets of X satisfying the following conditions:

(i) if $A \in \mathscr{F}$ and $B \in \mathscr{F}$, then $A \cap B \in \mathscr{F}$ (in particular, $A \cap B \neq \varnothing$);
(ii) if $A \in \mathscr{F}$ and if A′ is a subset of X containing A, then $A' \in \mathscr{F}$.

One calls *filter base* on X a set $\mathscr{B}$ of nonempty subsets of X satisfying the following condition:

(i′) if $A \in \mathscr{B}$ and $B \in \mathscr{B}$, there exists $C \in \mathscr{B}$ such that $C \subset A \cap B$ (in particular, $A \cap B \neq \varnothing$).

A filter is a filter base, but the converse is not true. If $\mathscr{B}$ is a filter base on X, one sees immediately that the set of subsets of X that contain an element of $\mathscr{B}$ is a filter.

2.1.2. Example. Let X be a topological space, $x_0 \in X$. The set $\mathscr{V}$ of neighborhoods of x_0 is a filter on X (1.3.4). If $\mathscr{W}$ is a fundamental system of neighborhoods of x_0, then $\mathscr{W}$ is a filter base on X.

2.1.3. Example. Let $x_0 \in \mathbf{R}$. The set of intervals $(x_0 - \varepsilon, x_0 + \varepsilon)$, where $\varepsilon > 0$, is a filter base on **R**. This is, moreover, a special case of 2.1.2. But here are some examples of filter bases on **R** that are not special cases of 2.1.2:

the set of $[x_0, x_0 + \varepsilon)$, where $\varepsilon > 0$;

the set of $(x_0, x_0 + \varepsilon)$, where $\varepsilon > 0$;

the set of $(x_0 - \varepsilon, x_0]$, where $\varepsilon > 0$;

the set of $(x_0 - \varepsilon, x_0)$, where $\varepsilon > 0$;

the set of $(x_0 - \varepsilon, x_0) \cup (x_0, x_0 + \varepsilon)$, where $\varepsilon > 0$.

2.1.4. Example. The set of intervals $[a, +\infty)$, where $a \in \mathbf{R}$, is a filter base on **R**. Similarly for the set of $(-\infty, a]$.

2.1.5. Example. On **N**, the set of subsets $\{n, n + 1, n + 2, \ldots\}$, where $n \in \mathbf{N}$, is a filter base.

2.1.6. Example. Let X be a topological space, $Y \subset X$, and $x_0 \in \overline{Y}$. The set of subsets of Y of the form $Y \cap V$, where V is a neighborhood of x_0 in X, is a filter on Y (notably $Y \cap V \neq \varnothing$ because $x_0 \in \overline{Y}$). If $Y = X$, one recovers 2.1.2.

2.2. Limits Along a Filter Base

2.2.1. Definition. Let X be a set equipped with a filter base $\mathscr{B}$, E a topological space, f a mapping of X into E, l a point of E. One says that *f tends to l along $\mathscr{B}$* if the following condition is satisfied:

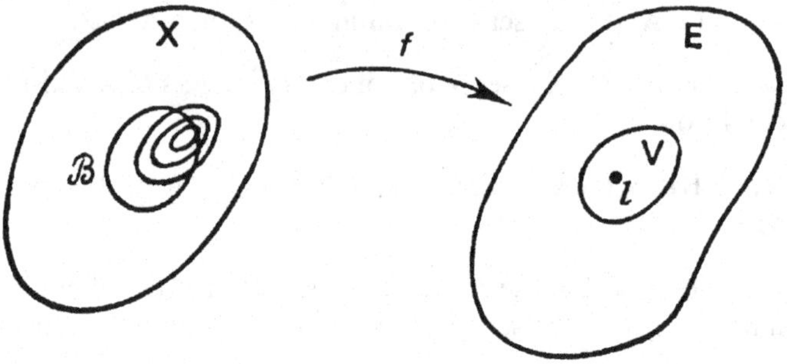

for every neighborhood V of l in E, there exists $B \in \mathscr{B}$ such that $f(B) \subset V$.

If one knows a fundamental system (V_i) of neighborhoods of l in E, it suffices to verify this condition for the V_i (for, every neighborhood of l contains a V_i).

2.2.2. Example. Suppose X = N, with the filter base $\mathscr{B}$ considered in 2.1.5. A mapping of N into E is nothing more than a sequence $(a_0, a_1, a_2, \ldots)$ of points of E. To say that this sequence tends to l along $\mathscr{B}$ means:

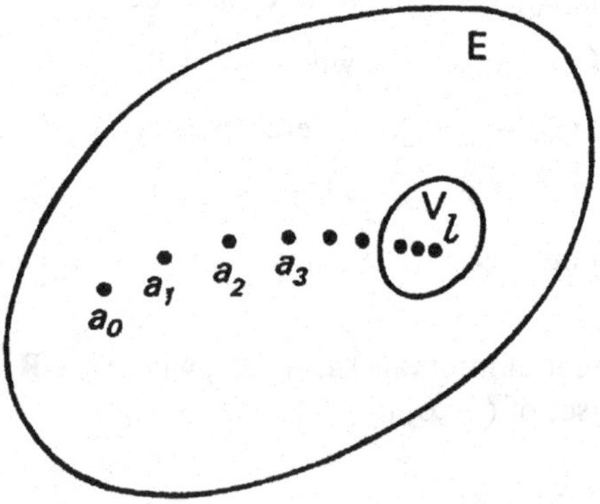

for every neighborhood V of l in E, there exists a positive integer N such that $n \geq N \Rightarrow a_n \in V$.

One then writes $\lim_{n \to \infty} a_n = l$.

If, for example, E is a metric space, this condition can be interpreted as follows:

for every $\varepsilon > 0$, there exists N such that $n \geq N \Rightarrow d(a_n, l) \leq \varepsilon$.

One recognizes the classical definition of the limit of a sequence of points in a metric space (for example, of a sequence of real numbers).

2.2.3. Example. Let X and E be topological spaces, f a mapping of X into E, $x_0 \in X$, $l \in E$. Take for $\mathscr{B}$ the filter of neighborhoods of x_0 in X (2.1.2). To say that f tends to l along $\mathscr{B}$ means:

for every neighborhood V of l in E, there exists a neighborhood W of x_0 in X such that $x \in W \Rightarrow f(x) \in V$.

One then writes $\lim_{x \to x_0} f(x) = l$.

If $X = E = \mathbf{R}$ one recovers the definition of the limit of a real-valued function of a real variable at a point.

2.2.4. Examples. One knows that the concept of limit of a real-valued function of a real variable admits many variants. These variants fit into the general framework of 2.2.1. For example, if $X = E = \mathbf{R}$ and if one takes the filter bases considered in 2.1.3, 2.1.4, one recovers the following known concepts:

$$\lim_{x \to x_0,\, x \geq x_0} f(x)$$

$$\lim_{x \to x_0,\, x > x_0} f(x)$$

$$\lim_{x \to x_0,\, x \leq x_0} f(x)$$

$$\lim_{x \to x_0,\, x < x_0} f(x)$$

$$\lim_{x \to x_0,\, x \neq x_0} f(x)$$

$$\lim_{x \to +\infty} f(x)$$

$$\lim_{x \to -\infty} f(x).$$

2.2.5. Example. Let X and E be topological spaces, $Y \subset X$, $x_0 \in \overline{Y}$, f a mapping of Y into E, $l \in E$. Take for $\mathscr{B}$ the filter defined in 2.1.6. To say that f tends to l along $\mathscr{B}$ means:

for every neighborhood V of l in E, there exists a neighborhood W of x_0 in X such that $x \in Y \cap W \Rightarrow f(x) \in V$.

One then writes $\lim_{x \to x_0,\, x \in Y} f(x) = l$.

This example generalizes 2.2.3. On taking $X = E = \mathbf{R}$, and for Y various subsets of $\mathbf{R}$, one recovers the first five examples of 2.2.4.

2.2.6. Theorem. *Let* X, E *be topological spaces, f a mapping of* X *into* E, $x_0 \in X$, $l \in E$, $(W_i)_{i \in I}$ *a fundamental system of neighborhoods of* x_0 *in* X, $(V_j)_{j \in J}$ *a fundamental system of neighborhoods of l in* E. *The following conditions are equivalent*:

(i) $\lim_{x \to x_0} f(x) = l$;
(ii) *for every* $j \in J$, *there exists* $i \in I$ *such that* $f(W_i) \subset V_j$.

Suppose condition (i) is satisfied. Let $j \in J$. There exists a neighborhood W of x_0 in X such that $f(W) \subset V_j$. Next, there exists $i \in I$ such that $W_i \subset W$. Then $f(W_i) \subset V_j$.

Suppose condition (ii) is satisfied. Let V be a neighborhood of l in E. There exists $j \in J$ such that $V_j \subset V$. Then there exists $i \in I$ such that $f(W_i) \subset V_j$, therefore $f(W_i) \subset V$.

2.2.7. Corollary. *Let* X, E *be metric spaces, f a mapping of* X *into* E, $x_0 \in X$, $l \in E$. *The following conditions are equivalent*:

(i) $\lim_{x \to x_0} f(x) = l$;
(ii) *for every* $\varepsilon > 0$, *there exists* $\eta > 0$ *such that*

$$x \in X,\ d(x, x_0) \leq \eta \ \Rightarrow\ d(f(x), l) \leq \varepsilon.$$

For, the closed balls with center x_0 (resp. l) and radius > 0 in X (resp. E) form a fundamental system of neighborhoods of x_0 (resp. l) in X (resp. E).

2.2.8. Theorem. *Let* X *be a set equipped with a filter base* $\mathscr{B}$, E *a separated topological space, f a mapping of* X *into* E. *If f admits a limit along* $\mathscr{B}$, *this limit is unique.*

Let l, l' be distinct limits of f along $\mathscr{B}$. Since E is separated, there exist disjoint neighborhoods V, V′ of l, l' in E. There exist $B, B' \in \mathscr{B}$ such that $f(B) \subset V$, $f(B') \subset V'$. Next, there exists $B'' \in \mathscr{B}$ such that $B'' \subset B \cap B'$. Then $f(B'') \subset V \cap V'$. Since $B'' \neq \varnothing$, one has $f(B'') \neq \varnothing$, therefore $V \cap V' \neq \varnothing$, which is absurd.

2.2.9. However, if E is not separated, f may admit more than one limit along $\mathscr{B}$. For example, if E is a coarse space, one verifies easily that every point of E is a limit of f along $\mathscr{B}$. The use of the limit concept in non-separated spaces is risky; in this course, we shall scarcely speak of limits except for separated spaces. Under these qualifications, Theorem 2.2.8 permits one to speak of *the* limit (if it exists! the reader already knows many examples of mappings that have no limit at all).

2.2.10. Theorem (Local Character of the Limit). *Let* X *be a set equipped with a filter base* $\mathscr{B}$, E *a topological space,* f *a mapping of* X *into* E, $l \in$ E. *Let* $X' \in \mathscr{B}$ *and let* f' *be the restriction of* f *to* X'. *The sets* $B \cap X'$, *where* $B \in \mathscr{B}$, *form a filter base* $\mathscr{B}'$ *on* X'. *The following conditions are equivalent*:

(i) f *tends to* l *along* $\mathscr{B}$;
(ii) f' *tends to* l *along* $\mathscr{B}'$.

Suppose that f tends to l along $\mathscr{B}$. Let V be a neighborhood of l in E. There exists $B \in \mathscr{B}$ such that $f(B) \subset V$. Then $f'(B \cap X') \subset V$ and $B \cap X' \in \mathscr{B}'$, thus f' tends to l along $\mathscr{B}'$.

Suppose that f' tends to l along $\mathscr{B}'$. Let V be a neighborhood of l in E. There exists $B' \in \mathscr{B}'$ such that $f'(B') \subset V$. But B' is of the form $B \cap X'$ with $B \in \mathscr{B}$. Since $X' \in \mathscr{B}$, there exists $B_1 \in \mathscr{B}$ such that $B_1 \subset B \cap X'$. Then $f(B_1) \subset f'(B') \subset V$, thus f tends to l along $\mathscr{B}$.

2.3. Mappings Continuous at a Point

2.3.1. Definition. Let X and Y be topological spaces, f a mapping of X into Y, and $x_0 \in X$. One says that f is *continuous at* x_0 if $\lim_{x \to x_0} f(x) = f(x_0)$, in other words (2.2.3) if the following condition is satisfied:

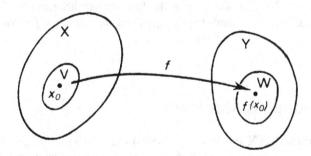

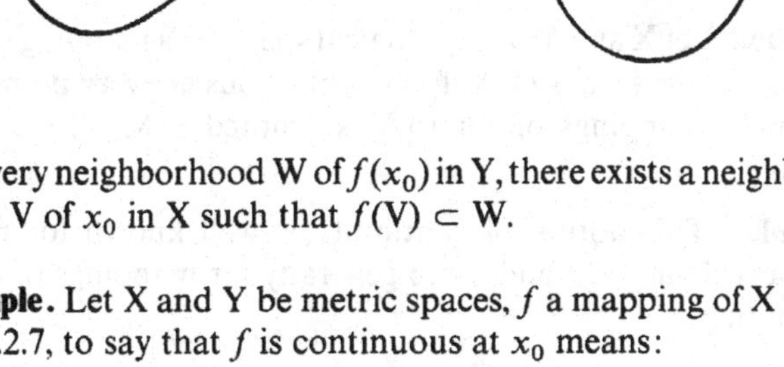

for every neighborhood W of $f(x_0)$ in Y, there exists a neighborhood V of x_0 in X such that $f(V) \subset W$.

2.3.2. Example. Let X and Y be metric spaces, f a mapping of X into Y, and $x_0 \in X$. By 2.2.7, to say that f is continuous at x_0 means:

for every $\varepsilon > 0$, there exists $\eta > 0$ such that
$x \in X$ and $d(x, x_0) \leq \eta \Rightarrow d(f(x), f(x_0)) \leq \varepsilon$.

We recognize here a classical definition. For example, if $X = Y = \mathbf{R}$, one recovers the continuity of a real-valued function of a real variable at a point.

2.3.3. Theorem. *Let* T *be a set equipped with a filter base* $\mathscr{B}$, X *and* Y *topological spaces,* $l \in$ X, f *a mapping of* T *into* X *that tends to* l *along* $\mathscr{B}$, *and* g *a mapping of* X *into* Y *that is continuous at* l. *Then* $g \circ f$ *tends to* $g(l)$ *along* $\mathscr{B}$.

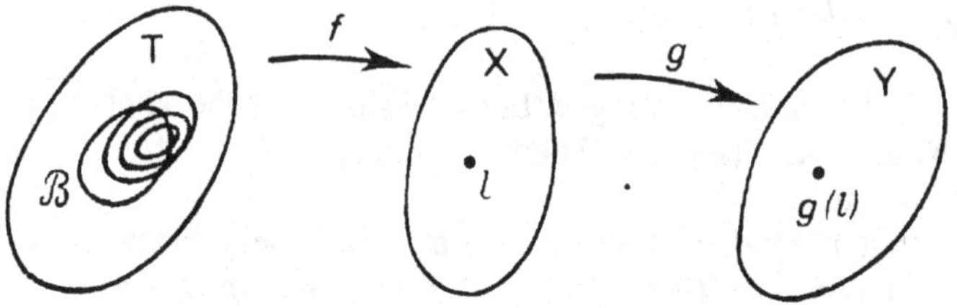

Let W be a neighborhood of $g(l)$ in Y. There exists a neighborhood V of l in X such that $g(\mathrm{V}) \subset \mathrm{W}$. Next, there exists a $\mathrm{B} \in \mathscr{B}$ such that $f(\mathrm{B}) \subset \mathrm{V}$. Then $(g \circ f)(\mathrm{B}) \subset g(\mathrm{V}) \subset \mathrm{W}$, whence the theorem.

2.3.4. Corollary. *Let* T, X, Y *be topological spaces,* $f: \mathrm{T} \to \mathrm{X}$ *and* $g: \mathrm{X} \to \mathrm{Y}$ *mappings, and* $t_0 \in \mathrm{T}$. *If* f *is continuous at* t_0, *and* g *is continuous at* $f(t_0)$, *then* $g \circ f$ *is continuous at* t_0.

One applies 2.3.3 on taking for $\mathscr{B}$ the filter of neighborhoods of t_0 in T, and $l = f(t_0)$. Then $g \circ f$ tends to $g(f(t_0))$ along $\mathscr{B}$, that is, $g \circ f$ is continuous at t_0.

2.4. Continuous Mappings

2.4.1. Definition. Let X and Y be topological spaces, f a mapping of X into Y. One says that f is *continuous on* X if f is continuous at every point of X. The set of continuous mappings of X into Y is denoted $\mathscr{C}(\mathrm{X}, \mathrm{Y})$.

2.4.2. Examples. This notion of continuity is well known for real-valued functions of a real variable, and more generally for mappings of one metric space into another.

2.4.3. Theorem. *Let* X, Y, Z *be topological spaces,* $f \in \mathscr{C}(\mathrm{X}, \mathrm{Y})$ *and* $g \in \mathscr{C}(\mathrm{Y}, \mathrm{Z})$. *Then* $g \circ f \in \mathscr{C}(\mathrm{X}, \mathrm{Z})$.

This follows at once from 2.3.4.

▶ **2.4.4. Theorem.** *Let X and Y be topological spaces, f a mapping of X into Y. The following conditions are equivalent:*

(i) *f is continuous;*
(ii) *the inverse image under f of every open subset of Y is an open subset of X;*
(iii) *the inverse image under f of every closed subset of Y is a closed subset of X;*
(iv) *for every subset A of X, $f(\overline{A}) \subset \overline{f(A)}$.*

(i) ⇒ (iv). Suppose f is continuous. Let $A \subset X$ and $x_0 \in \overline{A}$. Let W be a neighborhood of $f(x_0)$ in Y. Since f is continuous at x_0, there exists a neighborhood V of x_0 in X such that $f(V) \subset W$. Since $x_0 \in \overline{A}$, $V \cap A \neq \varnothing$. Since $f(V \cap A) \subset W \cap f(A)$, one sees that $W \cap f(A) \neq \varnothing$. This being true for every neighborhood W of $f(x_0)$, one has $f(x_0) \in \overline{f(A)}$. Thus $f(\overline{A}) \subset \overline{f(A)}$.

(iv) ⇒ (iii). Suppose condition (iv) is satisfied. Let Y′ be a closed subset of Y and let $X' = f^{-1}(Y')$. Then $f(X') \subset Y'$, therefore $\overline{f(X')} \subset Y'$ (1.5.4). If $x \in \overline{X'}$ then $f(x) \in \overline{f(X')}$ by condition (iv), therefore $f(x) \in Y'$ and so $x \in X'$. Thus $X' = \overline{X'}$, which proves that X′ is closed.

(iii) ⇒ (ii). Suppose condition (iii) is satisfied. Let Y′ be an open subset of Y. Then $Y - Y'$ is closed, therefore $f^{-1}(Y - Y')$ is closed. But $f^{-1}(Y - Y') = X - f^{-1}(Y')$. Therefore $f^{-1}(Y')$ is open.

(ii) ⇒ (i). Suppose condition (ii) is satisfied. Let $x_0 \in X$; let us prove that f is continuous at x_0. Let W be a neighborhood of $f(x_0)$ in Y. There exists an open subset Y′ of Y such that $f(x_0) \in Y' \subset W$. Let $X' = f^{-1}(Y')$. Then X′ is open by condition (ii), and $x_0 \in X'$, thus X′ is a neighborhood of x_0. Since $f(X') \subset Y' \subset W$, this proves the continuity of f at x_0.

2.4.5. Example. Let a, b be numbers > 0. One knows that the mapping $(x, y) \mapsto x^2/a^2 + y^2/b^2 - 1$ of $\mathbf{R}^2$ into $\mathbf{R}$ is continuous. On the other hand, $[0, +\infty)$ is a closed subset of $\mathbf{R}$. Therefore the set of $(x, y) \in \mathbf{R}^2$ such that $x^2/a^2 + y^2/b^2 - 1 \geq 0$ is closed in $\mathbf{R}^2$. Similarly, the set of $(x, y) \in \mathbf{R}^2$ such that $x^2/a^2 + y^2/b^2 - 1 = 0$ (an ellipse) is closed in $\mathbf{R}^2$, etc.

2.4.6. Mistake to Avoid. There is a risk of confusing conditions (ii) and (iii) of 2.4.4 with the following conditions:

(ii′) the direct image of every open subset of X is an open subset of Y;
(iii′) the direct image of every closed subset of X is a closed subset of Y.

Mappings satisfying (ii′) (resp. (iii′)) are called open mappings (resp. closed mappings). There exist continuous mappings that are neither open nor closed, open mappings that are neither continuous nor closed, and closed mappings that are neither continuous nor open.

2.4.7. Let $\mathcal{T}_1$, $\mathcal{T}_2$ be two topologies on a set E. Denote by E_1, E_2 the set E equipped with the topologies $\mathcal{T}_1$, $\mathcal{T}_2$. To say that $\mathcal{T}_1$ is finer than $\mathcal{T}_2$ means, by 2.4.4(ii), that the identity mapping of E_1 into E_2 is continuous.

2.5. Homeomorphisms

2.5.1. Theorem. *Let* X *and* Y *be topological spaces, f a bijective mapping of* X *onto* Y. *The following conditions are equivalent*:

(i) f *and* f^{-1} *are continuous*;
(ii) *in order that a subset of* X *be open, it is necessary and sufficient that its image in* Y *be open*;
(iii) *in order that a subset of* X *be closed, it is necessary and sufficient that its image in* Y *be closed.*

This follows at once from 2.4.4.

2.5.2. Definition. A mapping f of X into Y that satisfies the conditions of 2.5.1 is called a *bicontinuous* mapping of X onto Y, or a *homeomorphism* of X onto Y. (By 2.5.1(ii), this is the natural concept of isomorphism for the structure of topological space.)

2.5.3. It is clear that the inverse mapping of a homeomorphism is a homeomorphism. By 2.4.3, the composite of two homeomorphisms is a homeomorphism.

2.5.4. Let X and Y be topological spaces. If there exists a homeomorphism of X onto Y, then X and Y are said to be *homeomorphic*. By virtue of 2.5.3, this is an equivalence relation among topological spaces. If X and Y are homeomorphic, the open set structure is the same in X and Y; since all topological properties are defined on the basis of open sets, X and Y will have the same topological properties; X and Y are almost the 'same' topological space. One of the goals of topology (not the only one, far from it) consists in recognizing whether or not two given spaces are homeomorphic, and classifying topological spaces up to homeomorphism; this goal is far from being attained at the present time.

2.5.5. Examples. All nonempty open intervals in **R** are homeomorphic. For, if the open intervals I_1 and I_2 are bounded, there exists a homothety or a translation f that transforms I_1 into I_2, and f is obviously bicontinuous. Similarly if I_1 and I_2 are of the form $(a, +\infty)$ or $(-\infty, a)$. Thus, it remains to compare the intervals $(0, 1)$, $(0, +\infty)$ and $(-\infty, +\infty)$. Now, the mapping $x \mapsto \tan(\pi/2)x$ of $(0, 1)$ into $(0, +\infty)$ is bijective and continuous, and the inverse mapping $x \mapsto (2/\pi)\text{Arctan}\, x$ is continuous; therefore $(0, 1)$ and $(0, +\infty)$ are homeomorphic. Similarly, the mapping $x \mapsto \tan(\pi/2)x$ of $(-1, 1)$ into $(-\infty, +\infty)$ is a homeomorphism.

The intervals $(0, 1)$ and $[0, 1]$ are *not* homeomorphic (cf. 4.2.8).

2.5.6. Example. A circle and a square in $\mathbf{R}^2$ are homeomorphic (via a translation followed by a central projection).

2.5.7. Example: Stereographic Projection. Let $\mathbf{S}_n$ be the set of

$$x = (x_1, x_2, \ldots, x_{n+1}) \in \mathbf{R}^{n+1}$$

such that $x_1^2 + \cdots + x_{n+1}^2 = 1$ ('n-dimensional sphere'). Let

$$a = (0, 0, \ldots, 0, 1) \in \mathbf{S}_n.$$

Let us identify $\mathbf{R}^n$ with the set of $(x_1, \ldots, x_n, 0) \in \mathbf{R}^{n+1}$. We are going to define a homeomorphism of $\mathbf{S}_n - \{a\}$ onto $\mathbf{R}^n$.

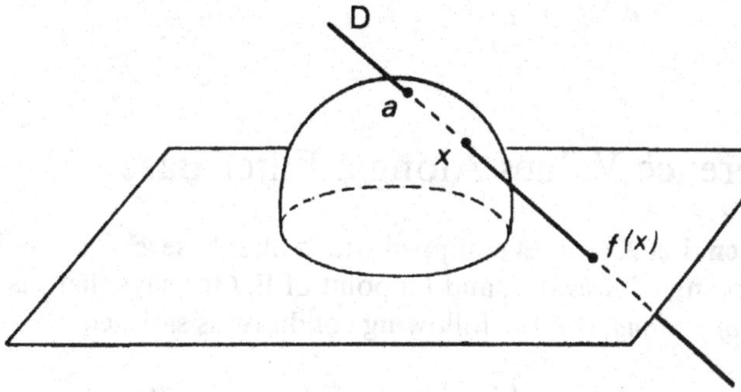

Let $x = (x_1, \ldots, x_{n+1}) \in \mathbf{S}_n - \{a\}$. The line D in $\mathbf{R}^{n+1}$ joining a and x is the set of points of the form

$$(\lambda x_1, \ldots, \lambda x_n, 1 + \lambda(x_{n+1} - 1))$$

with $\lambda \in \mathbf{R}$. This point is in $\mathbf{R}^n$ when $1 + \lambda(x_{n+1} - 1) = 0$, that is, when $\lambda = (1 - x_{n+1})^{-1}$ ($x_{n+1} \neq 1$ because $x \neq a$). Thus $\mathrm{D} \cap \mathbf{R}^n$ reduces to the point $f(x)$ with coordinates

$$(1) \qquad x_1' = \frac{x_1}{1 - x_{n+1}}, \qquad x_2' = \frac{x_2}{1 - x_{n+1}}, \ldots, x_n' = \frac{x_n}{1 - x_{n+1}}, \qquad x_{n+1}' = 0.$$

We have thus defined a mapping f of $\mathbf{S}_n - \{a\}$ into $\mathbf{R}^n$. Given $x' = (x_1', \ldots, x_n', 0)$ in $\mathbf{R}^n$, there exists one and only one point

$$x = (x_1, \ldots, x_{n+1}) \in \mathbf{S}_n - \{a\}$$

such that $f(x) = x'$. For, the solution of (1) yields the conditions

$$x_i = x_i'(1 - x_{n+1}) \quad \text{for} \quad 1 \leq i \leq n,$$

$$\sum_{i=1}^{n} x_i'^2(1 - x_{n+1})^2 + x_{n+1}^2 = 1,$$

then, after dividing out the nonzero factor $1 - x_{n+1}$,

$$(x_1'^2 + \cdots + x_n'^2)(1 - x_{n+1}) - 1 - x_{n+1} = 0,$$

whence

$$(2)\qquad \begin{cases} x_{n+1} = \dfrac{x_1'^2 + \cdots + x_n'^2 - 1}{x_1'^2 + \cdots + x_n'^2 + 1}, \\[2ex] x_i = \dfrac{2x_i'}{x_1'^2 + \cdots + x_n'^2 + 1} \qquad (1 \le i \le n). \end{cases}$$

Thus f is a bijection of $\mathbf{S}_n - \{a\}$ onto $\mathbf{R}^n$. The formulas (1) and (2) prove, moreover, that f and f^{-1} are continuous.

The homeomorphism f is called stereographic projection of $\mathbf{S}_n - \{a\}$ onto $\mathbf{R}^n$.

2.6. Adherence Values Along a Filter Base

2.6.1. Definition. Let X be a set equipped with a filter base $\mathscr{B}$, E a topological space, f a mapping of X into E, and l a point of E. One says that l is an *adherence value of f along $\mathscr{B}$* if the following condition is satisfied:

for every neighborhood V of l in E and for every $\mathrm{B} \in \mathscr{B}$, $f(\mathrm{B})$ intersects V.

If one knows a fundamental system of neighborhoods (V_i) of l in E, it suffices to verify this condition for the V_i.

2.6.2. Example. Suppose in 2.6.1 that $\mathrm{X} = \mathbf{N}$, with the filter base 2.1.5. We are thus considering a sequence $(a_0, a_1, \ldots)$ of points of E. To say that l is an adherence value of the sequence means:

for every neighborhood V of l in E, and every positive integer N, there exists $n \ge \mathrm{N}$ such that $a_n \in \mathrm{V}$.

If, for example, E is a metric space, this condition may be rewritten as follows:

for every $\varepsilon > 0$, and every positive integer N, there exists $n \ge \mathrm{N}$ such that $d(a_n, l) \le \varepsilon$.

For example, taking $\mathrm{E} = \mathbf{R}$, consider the sequence of numbers

$$\tfrac{1}{4},\quad 1 - \tfrac{1}{4},\quad \tfrac{1}{5},\quad 1 - \tfrac{1}{5},\quad \tfrac{1}{6},\quad 1 - \tfrac{1}{6}, \ldots.$$

The verification that the adherence values of this sequence are 0 and 1 is left as an exercise.

2.6.3. Example. Let X and E be topological spaces, $Y \subset X$, $x_0 \in \overline{Y}$, f a mapping of Y into E, and $l \in E$. Take for $\mathscr{B}$ the filter 2.1.6. To say that l is an adherence value of f along $\mathscr{B}$ means:

for every neighborhood V of l in E, and every neighborhood W of x_0 in X, $f(W \cap Y)$ intersects V.

One then says that l is an adherence value of f as x tends to x_0 while remaining in Y.

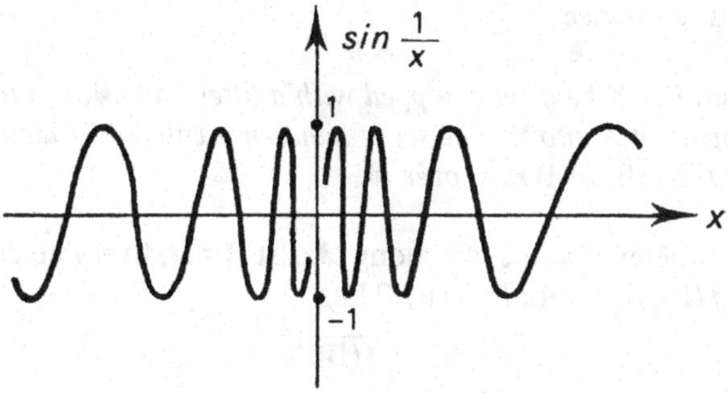

For example, take $X = E = \mathbf{R}$, $Y = \mathbf{R} - \{0\}$, $x_0 = 0$, and $f(x) = \sin(1/x)$ for $x \in \mathbf{R} - \{0\}$. It is left as an exercise to show that the adherence values of f, as x tends to 0 through values $\neq 0$, are all the numbers in $[-1, 1]$.

From the examples 2.6.2, 2.6.3, one recognizes that the concept of adherence value is a kind of substitute for the limit concept. This point will now be elaborated.

2.6.4. Theorem. *Let* X *be a set equipped with a filter base* $\mathscr{B}$, E *a separated topological space, f a mapping of* X *into* E, *and l a point of* E. *If f tends to l along* $\mathscr{B}$, *then l is the unique adherence value of f along* $\mathscr{B}$.

Let V be a neighborhood of l in E and let $B \in \mathscr{B}$. There exists $B' \in \mathscr{B}$ such that $f(B') \subset V$. Then $B \cap B' \neq \varnothing$, therefore $f(B \cap B') \neq \varnothing$, and $f(B \cap B') \subset f(B) \cap V$. Therefore $f(B)$ intersects V. Thus l is an adherence value of f along $\mathscr{B}$.

Let l' be an adherence value of f along $\mathscr{B}$, and suppose $l' \neq l$. There exist neighborhoods V, V′ of l, l' that are disjoint. Next, there exists $B \in \mathscr{B}$ such that $f(B) \subset V$. Then $f(B) \cap V' = \varnothing$, which contradicts the fact that l' is an adherence value.

2.6.5. Thus, when f admits a limit along $\mathscr{B}$, the concept of adherence value brings nothing new with it and is thus without interest. But what can happen if f does *not* have a limit along $\mathscr{B}$?

(a) it can happen that f has no adherence value; example: the sequence $(0, 1, 2, 3, \ldots)$ in $\mathbf{R}$ has no adherence value; however, cf. 4.2.1;
(b) it can happen that f has a unique adherence value; example: the sequence $(0, 1, 0, 2, 0, 3, \ldots)$ in $\mathbf{R}$ has no limit, and its only adherence value is 0; however, cf. 4.2.4;
(c) it can happen that f has more than one adherence value (cf. 2.6.2, 2.6.3).

To sum up, in relation to the limit concept, one loses uniqueness but one gains as regards existence.

2.6.6. Theorem. *Let* X *be a set equipped with a filter base* $\mathscr{B}$, E *a topological space,* f *a mapping of* X *into* E. *The set of adherence values of* f *along* $\mathscr{B}$ *is the intersection of the* $\overline{f(\mathrm{B})}$ *as* B *runs over* $\mathscr{B}$.

Let l be an adherence value of f along $\mathscr{B}$. Let $\mathrm{B} \in \mathscr{B}$. Every neighborhood of l intersects $f(\mathrm{B})$, therefore $l \in \overline{f(\mathrm{B})}$. Thus

$$l \in \bigcap_{\mathrm{B} \in \mathscr{B}} \overline{f(\mathrm{B})}.$$

Suppose $m \in \bigcap_{\mathrm{B} \in \mathscr{B}} \overline{f(\mathrm{B})}$. Let V be a neighborhood of m and let $\mathrm{B} \in \mathscr{B}$. Since $m \in \overline{f(\mathrm{B})}$, $f(\mathrm{B})$ intersects V. Therefore m is an adherence value of f.

(There is thus a connection between the concept of adherence value and that of adherent point. However, the two concepts should not be confused.)

CHAPTER III

Constructions of Topological Spaces

The study of any structure often leads to the study of substructures, product structures and quotient structures. For example, the student has already seen this in the study of vector space structure. The same is true for topological spaces. This yields important new spaces (for example, the tori $\mathbf{T}^n$; cf. also the projective spaces, in the exercises for Chapter IV).

3.1. Topological Subspaces

3.1.1. Theorem. *Let* E *be a topological space,* F *a subset of* E. *Let* $\mathscr{U}$ *be the set of open subsets of* E. *Let* $\mathscr{V}$ *be the set of subsets of* F *of the form* $U \cap F$, *where* $U \in \mathscr{U}$. *Then* $\mathscr{V}$ *satisfies the axioms* (i), (ii), (iii) *of* 1.2.1.

(i) One has $\varnothing \in \mathscr{U}$ and $E \in \mathscr{U}$, therefore $\varnothing = \varnothing \cap F \in \mathscr{V}$ and $F = E \cap F \in \mathscr{V}$.

(ii) Let $(V_i)_{i \in I}$ be a family of subsets belonging to $\mathscr{V}$. For every $i \in I$, there exists $U_i \in \mathscr{U}$ such that $V_i = U_i \cap F$. Then $\bigcup_{i \in I} U_i \in \mathscr{U}$, therefore

$$\bigcup_{i \in I} V_i = \bigcup_{i \in I} (U_i \cap F) = \left(\bigcup_{i \in I} U_i\right) \cap F \in \mathscr{V}.$$

(iii) With notations as in (ii), suppose moreover that I is finite. Then $\bigcap_{i \in I} U_i \in \mathscr{U}$, therefore

$$\bigcap_{i \in I} V_i = \bigcap_{i \in I} (U_i \cap F) = \left(\bigcap_{i \in I} U_i\right) \cap F \in \mathscr{V}.$$

3.1.2. By 3.1.1, $\mathscr{V}$ is the set of open sets of a topology on F, called the *topology induced on* F by the given topology of E. Equipped with this topology, F is called a *topological subspace* (or simply a subspace) of E. *The open sets of* F *are thus, by definition, the intersections with* F *of the open sets of* E.

3.1.3. Remark. Let E be a topological space, F a subspace of E. If A is a subset of F, the properties of A relative to F and relative to E may differ. For example, if A is open in E then A is open in F (because $A = A \cap F$), but the converse is in general not true (for example, F is an open subset of F but in general is not an open subset of E). However, if F is an open subset of E and if A is open in F, then A is open in E (because $A = B \cap F$ with B open in E, and the intersection of two open subsets of E is an open subset of E).

3.1.4. Theorem. *Let* E *be a topological space,* F *a subspace of* E, A *a subset of* F. *The following conditions are equivalent:*

(i) A *is closed in* F;
(ii) A *is the intersection with* F *of a closed subset of* E.

(i) $\Rightarrow$ (ii). Suppose A is closed in F. Then $F - A$ is open in F, therefore there exists an open subset U of E such that $F - A = U \cap F$. Since $A = (E - U) \cap F$ and since $E - U$ is closed in E, we see that condition (ii) is satisfied.

(ii) $\Rightarrow$ (i). Suppose $A = X \cap F$, where X is a closed subset of E. Then $F - A = (E - X) \cap F$, and $E - X$ is open in E, therefore $F - A$ is open in F, thus A is closed in F.

3.1.5. Remark. Let us maintain the notations of 3.1.4. If A is closed in E then A is closed in F, but the converse is in general not true (for example, F is closed in F but in general not in E). However, if F is closed in E and if A is closed in F, then A is closed in E (because $A = X \cap F$ with X closed in E, and the intersection of two closed subsets of E is a closed subset of E).

3.1.6. Theorem. *Let* E *be a topological space,* F *a subspace of* E, *and* $x \in F$. *Let* $W \subset F$. *The following conditions are equivalent:*

(i) W *is a neighborhood of* x *in* F;
(ii) W *is the intersection with* F *of a neighborhood of* x *in* E.

(i) $\Rightarrow$ (ii). Suppose W is a neighborhood of x in F. There exists an open subset B of F such that $x \in B \subset W$. Then there exists an open subset A of E such that $B = F \cap A$. Let $V = A \cup W$. Then $x \in A \subset V$, thus V is a neighborhood of x in E. On the other hand,

$$F \cap V = (F \cap A) \cup (F \cap W) = B \cup W = W.$$

(ii) ⇒ (i). Suppose $W = F \cap V$, where V is a neighborhood of x in E. There exists an open subset A of E such that $x \in A \subset V$. Then $x \in F \cap A \subset F \cap V = W$, and $F \cap A$ is open in F, thus W is a neighborhood of x in F.

3.1.7. Theorem. *Let* E *be a topological space,* F *a subspace of* E. *If* E *is separated, then* F *is separated.*

Let x, y be distinct points of F. There exist disjoint neighborhoods V, W of x, y in E. Then $F \cap V$, $F \cap W$ are neighborhoods of x, y in F (3.1.6) and they are disjoint. Thus F is separated.

3.1.8. Definition. Let E be a topological space, $A \subset E$ and $x \in A$. One says that x is an *isolated* point of A if there exists a neighborhood V of x in E such that $V \cap A = \{x\}$.

3.1.9. Theorem. *Let* E *be a topological space and let* $F \subset E$. *The following conditions are equivalent:*

(i) *the topological space* F *is discrete;*
(ii) *every point of* F *is isolated.*

(i) ⇒ (ii). Suppose F is discrete. Let $x \in F$. Then $\{x\}$ is an open subset of F, therefore there exists an open subset U of E such that $\{x\} = U \cap F$. Since U is a neighborhood of x in E, we see that x is an isolated point of F.

(ii) ⇒ (i). Suppose that every point of F is isolated. Let $x \in F$. There exists a neighborhood V of x in E such that $V \cap F = \{x\}$. Passing to a subset of V, we can suppose that V is open in E. Then $\{x\}$ is open in F. Since every subset of F is the union of one-element subsets, every subset of F is open in F. Thus the topological space F is discrete.

3.1.10. Example. In **R**, consider the subset **Z**. If $n \in \mathbf{Z}$ then

$$\{n\} = \mathbf{Z} \cap (n - \tfrac{1}{2}, n + \tfrac{1}{2}),$$

therefore n is isolated in **Z**. Thus the topological subspace **Z** of **R** is discrete.

3.1.11. Theorem (Transitivity of Subspaces). *Let* E, E′, E″ *be sets such that* $E \supset E' \supset E''$. *Let* $\mathscr{T}$ *be a topology on* E, $\mathscr{T}'$ *the topology induced by* $\mathscr{T}$ *on* E′, $\mathscr{T}''$ *the topology induced by* $\mathscr{T}'$ *on* E″. *Then* $\mathscr{T}''$ *is the topology induced by* $\mathscr{T}$ *on* E″.

Let $\mathscr{T}''_1$ be the topology induced by $\mathscr{T}$ on E″.

Let $U'' \subset E''$ be an open set for $\mathscr{T}''$. There exists a subset U′ of E′, open for $\mathscr{T}'$, such that $U' \cap E'' = U''$. Next, there exists a subset U of E, open for $\mathscr{T}$, such that $U \cap E' = U'$. Then $U \cap E'' = U''$, thus U″ is open for $\mathscr{T}''_1$.

Let $V'' \subset E''$ be an open set for $\mathcal{T}''_1$. There exists a subset V of E, open for $\mathcal{T}$, such that $V \cap E'' = V''$. Set $V' = V \cap E'$. Then V' is open for $\mathcal{T}'$, and $V'' = V' \cap E''$, thus V'' is open for $\mathcal{T}''$.

3.1.12. Theorem. *Let* E *be a metric space,* E′ *a metric subspace of* E(1.1.1). *Let* $\mathcal{T}$, $\mathcal{T}'$ *be the topologies of* E, E′ *defined by their metrics* (1.2.2). *Then* $\mathcal{T}'$ *is nothing more than the topology induced by* $\mathcal{T}$ *on* E′.

Let $\mathcal{T}'_1$ be the topology induced by $\mathcal{T}$ on E′.

Let $U' \subset E'$ be an open set for $\mathcal{T}'$. For every $x \in U'$, there exists $\varepsilon_x > 0$ such that the open ball B'_x in E′ with center x and radius ε_x is contained in U′. Let B_x be the open ball in E with center x and radius ε_x. Then B_x is a neighborhood of x for $\mathcal{T}$, therefore B'_x is a neighborhood of x for $\mathcal{T}'_1$ (3.1.6). Thus U′ is a neighborhood of x for $\mathcal{T}'_1$. This being true for every $x \in U'$, we see that U′ is open in E′ for $\mathcal{T}'_1$ (1.3.5).

Let $V' \subset E'$ be an open set for $\mathcal{T}'_1$. There exists an open subset V of E such that $V' = V \cap E'$. For every $x \in V'$, there exists $\eta_x > 0$ such that the open ball C_x in E with center x and radius η_x is contained in V. Let C'_x be the open ball in E′ with center x and radius η_x. Then $C'_x = C_x \cap E' \subset V \cap E' = V'$. Therefore V′ is open for $\mathcal{T}'$.

3.1.13. For example, if a subset A of $\mathbf{R}^n$ is regarded as a topological subspace of $\mathbf{R}^n$, the topology of A is none other than the one considered in 1.2.2.

3.1.14. Theorem. *Let* X *be a set equipped with a filter base* $\mathcal{B}$, E *a topological space,* E′ *a subspace of* E, f *a mapping of* X *into* E′, l *a point of* E′. *The following conditions are equivalent*:

(i) f *tends to* l *along* $\mathcal{B}$ *relative to* E′;
(ii) f *tends to* l *along* $\mathcal{B}$ *relative to* E.

Suppose condition (i) is satisfied. Let V be a neighborhood of l in E. Then $V \cap E'$ is a neighborhood of l in E′ (3.1.6). There exists $B \in \mathcal{B}$ such that $f(B) \subset V \cap E'$. *A fortiori*, $f(B) \subset V$. Thus f tends to l along $\mathcal{B}$ relative to E.

Suppose condition (ii) is satisfied. Let V′ be a neighborhood of l in E′. There exists a neighborhood V of l in E such that $V \cap E' = V'$ (3.1.6). Then there exists $B \in \mathcal{B}$ such that $f(B) \subset V$. Since $f(X) \subset E'$, one has $f(B) \subset V \cap E' = V'$. Thus f tends to l along $\mathcal{B}$ relative to E′.

▶ **3.1.15. Theorem.** *Let* T, E *be topological spaces,* E′ *a subspace of* E, f *a mapping of* T *into* E′. *The following conditions are equivalent*:

(i) f *is continuous*;
(ii) f, *regarded as a mapping of* T *into* E, *is continuous.*

Indeed, for every $t_0 \in T$, the condition $\lim_{t \to t_0} f(t) = f(t_0)$ has the same meaning, according to 3.1.14, whether one considers f to have values in E′ or to have values in E.

3.1.16. Corollary. *Let* E *be a topological space,* E′ *a subspace of* E. *The identity mapping of* E′ *into* E *is continuous.*

For, the identity mapping of E′ into E′ is obviously continuous. One then applies 3.1.15.

3.2. Finite Products of Topological Spaces

3.2.1. Let $E_1, E_2, \ldots, E_n$ be topological spaces. We are going to define a natural topology on $E = E_1 \times E_2 \times \cdots \times E_n$.

Let us call *elementary open set* in E a subset of the form $U_1 \times U_2 \times \cdots \times U_n$, where U_i is an open subset of E_i. Let us call *open set* in E any union of elementary open sets. To justify this terminology, we are going to show that this family of subsets of E satisfies the axioms (i), (ii), (iii) of 1.2.1.

First of all,

$$E = E_1 \times E_2 \times \cdots \times E_n \quad \text{and} \quad \varnothing = \varnothing \times E_2 \times \cdots \times E_n$$

are open sets, even elementary open sets.

Axiom (ii) is obvious.

Finally, let A, B be open subsets of E and let us show that $A \cap B$ is an open subset of E. One has $A = \bigcup A_\lambda$, $B = \bigcup B_\mu$, where the A_λ and B_μ are elementary open sets. Then $A \cap B$ is the union of the $A_\lambda \cap B_\mu$ and it suffices to prove that, for fixed λ and μ, $A_\lambda \cap B_\mu$ is an elementary open set. Now,

$$A_\lambda = U_1 \times \cdots \times U_n, \qquad B_\mu = V_1 \times \cdots \times V_n,$$

where U_i, V_i are open subsets of E_i. It follows that

$$A_\lambda \cap B_\mu = (U_1 \cap V_1) \times \cdots \times (U_n \cap V_n);$$

since $U_i \cap V_i$ is an open subset of E_i, this completes the proof.

3.2.2. We have thus defined a topology on E, called the *product topology* of the given topologies on $E_1, \ldots, E_n$. We also say that E is the *product topological space* of the topological spaces $E_1, E_2, \ldots, E_n$.

3.2.3. Example (Product of Metric Spaces). Let $E_1, E_2, \ldots, E_n$ be metric spaces. Set

$$E = E_1 \times \cdots \times E_n.$$

If $x = (x_1, \ldots, x_n) \in E$ and $y = (y_1, \ldots, y_n) \in E$, set

$$d(x, y) = (d(x_1, y_1)^2 + \cdots + d(x_n, y_n)^2)^{1/2}.$$

One verifies as in the case of $\mathbf{R}^n$ that d is a metric on E. Thus, a product of metric spaces is automatically a metric space. Let $\mathscr{T}$ be the topology on E defined by this metric (1.2.2). In addition, $E_1, \ldots, E_n$ are topological spaces (1.2.2), thus by 3.2.2 there is defined a product topology $\mathscr{T}'$ on E. *Let us show that* $\mathscr{T} = \mathscr{T}'$.

Let U be a subset of E that is open for $\mathscr{T}'$. For every $x = (x_1, \ldots, x_n) \in U$, there exist open neighborhoods $U_1, \ldots, U_n$ of $x_1, \ldots, x_n$ in $E_1, \ldots, E_n$ such that $U_1 \times \cdots \times U_n \subset U$. There exists $\varepsilon_i > 0$ such that the open ball in E_i with center x_i and radius ε_i is contained in U_i. Let B be the open ball in E with center x and radius $\varepsilon = \inf(\varepsilon_1, \ldots, \varepsilon_n)$. If $y = (y_1, \ldots, y_n) \in B$ then $d(x, y) < \varepsilon$, therefore $d(x_i, y_i) < \varepsilon \leq \varepsilon_i$ for all i, thus $y_i \in U_i$, and so $y \in U$. Thus $B \subset U$. We have thus proved that U is open for $\mathscr{T}$.

Let V be a subset of E that is open for $\mathscr{T}$. For every $x = (x_1, \ldots, x_n) \in V$, there exists $\varepsilon > 0$ such that V contains the open ball with center x and radius ε. Let B_i be the open ball in E_i with center x_i and radius ε/n. If

$$y = (y_1, \ldots, y_n) \in B_1 \times \cdots \times B_n,$$

then

$$d(x, y)^2 = \sum_{i=1}^{n} d(x_i, y_i)^2 < \sum_{i=1}^{n} \frac{\varepsilon^2}{n^2} \leq \varepsilon^2,$$

therefore $y \in V$. Thus, the elementary open set $B_1 \times \cdots \times B_n$ is contained in V and contains x. Consequently, V is the union of elementary open sets, therefore is open for $\mathscr{T}'$.

In particular, if $\mathbf{R}^n$ is equipped with the topology defined by its usual metric (1.1.2), $\mathbf{R}^n$ appears as the product topological space $\mathbf{R} \times \mathbf{R} \times \cdots \times \mathbf{R}$.

3.2.4. Theorem. *Let* $E = E_1 \times \cdots \times E_n$ *be a product of topological spaces. Let* $x = (x_1, \ldots, x_n) \in E$. *The sets of the form* $V_1 \times \cdots \times V_n$, *where* V_i *is a neighborhood of* x_i *in* E_i, *constitute a fundamental system of neighborhoods of* x *in* E.

For $i = 1, \ldots, n$, let V_i be a neighborhood of x_i in E_i. There exists an open subset U_i of E_i such that $x_i \in U_i \subset V_i$. Then

$$x \in U_1 \times \cdots \times U_n \subset V_1 \times \cdots \times V_n$$

and $U_1 \times \cdots \times U_n$ is open in E, thus $V_1 \times \cdots \times V_n$ is a neighborhood of x in E.

Let V be a neighborhood of x in E. There exists an open subset U of E such that $x \in U \subset V$. The set U is the union of elementary open sets, therefore x belongs to one of these sets, say $U_1 \times \cdots \times U_n$, where U_i is an open subset of E_i. Then $x_i \in U_i$, thus U_i is a neighborhood of x_i in E_i, and $U_1 \times \cdots \times U_n \subset V$.

3.2.5. Theorem. *Let* $E = E_1 \times \cdots \times E_n$ *be a product of topological spaces. If each* E_i *is separated, then* E *is separated.*

Let $x = (x_1, \ldots, x_n)$ and $y = (y_1, \ldots, y_n)$ be two distinct points of E. One has $x_i \neq y_i$ for at least one i, for example $x_1 \neq y_1$. There exist disjoint neighborhoods V, W of x_1, y_1 in E_1. Then $V \times E_2 \times \cdots \times E_n$ and $W \times E_2 \times \cdots \times E_n$ are neighborhoods of x, y in E (3.2.4) and they are disjoint.

3.2.6. Theorem. *Let* X *be a set equipped with a filter base* $\mathscr{B}$, $E = E_1 \times E_2 \times \cdots \times E_n$ *a product of topological spaces,* $l = (l_1, \ldots, l_n) \in E$. *Let* f *be a mapping of* X *into* E, *thus of the form* $x \mapsto (f_1(x), \ldots, f_n(x))$, *where* f_i *is a mapping of* X *into* E_i. *Then the following conditions are equivalent*:

(i) f *tends to* l *along* $\mathscr{B}$;
(ii) *for* $i = 1, 2, \ldots, n$, f_i *tends to* l_i *along* $\mathscr{B}$.

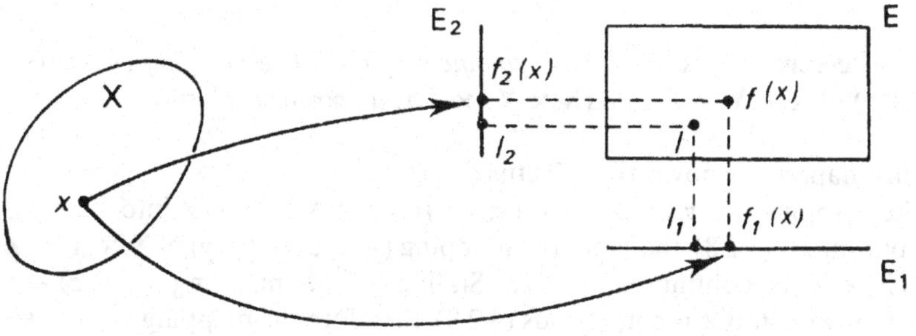

Suppose that f tends to l along $\mathscr{B}$. Let us show, for example, that f_1 tends to l_1 along $\mathscr{B}$. Let V_1 be a neighborhood of l_1 in E_1. Then $V_1 \times E_2 \times \cdots \times E_n$ is a neighborhood of l in E (3.2.4). Therefore there exists $B \in \mathscr{B}$ such that $f(B) \subset V_1 \times E_2 \times \cdots \times E_n$. Then $f_1(B) \subset V_1$, thus f_1 tends to l_1 along $\mathscr{B}$.

Suppose condition (ii) is satisfied. Let V be a neighborhood of l in E. There exist neighborhoods $V_1, \ldots, V_n$ of $l_1, \ldots, l_n$ in $E_1, \ldots, E_n$ such that $V_1 \times \cdots \times V_n \subset V$ (3.2.4). Then there exist $B_1, \ldots, B_n \in \mathscr{B}$ such that $f_1(B_1) \subset V_1, \ldots, f_n(B_n) \subset V_n$. Next, there exists $B \in \mathscr{B}$ such that $B \subset B_1 \cap \cdots \cap B_n$. Then

$$f(B) \subset f_1(B_1) \times \cdots \times f_n(B_n) \subset V_1 \times \cdots \times V_n \subset V,$$

thus f tends to l along $\mathscr{B}$.

▶ **3.2.7. Theorem.** *Let* $E = E_1 \times \cdots \times E_n$ *be a product of topological spaces, and* T *a topological space. Let* f *be a mapping of* T *into* E, *thus of the form* $t \mapsto (f_1(t), \ldots, f_n(t))$, *where* f_i *is a mapping of* T *into* E_i. *The following conditions are equivalent*:

(i) f *is continuous*;
(ii) $f_1, \ldots, f_n$ *are continuous.*

Indeed, for every $t_0 \in T$, the conditions

$$\lim_{t \to t_0} f(t) = f(t_0),$$

$$\lim_{t \to t_0} f_i(t) = f_i(t_0) \quad \text{for} \quad i = 1, \ldots, n$$

are equivalent by 3.2.6.

3.2.8. Corollary. *Let* $E = E_1 \times \cdots \times E_n$ *be a product of topological spaces. The canonical projections of* E *onto* $E_1, \ldots, E_n$ *are continuous.*

Let f be the identity mapping of E. It is continuous. Now, it is the mapping $x \mapsto (f_1(x), \ldots, f_n(x))$, where $f_1, \ldots, f_n$ are the canonical projections of E onto $E_1, \ldots, E_n$. It then suffices to apply 3.2.7.

3.2.9. Theorem. *Let* X, Y, Z *be topological spaces. The mapping* $(x, y, z) \mapsto ((x, y), z)$ *of* $X \times Y \times Z$ *onto* $(X \times Y) \times Z$ *is a homeomorphism.*

This mapping is obviously bijective.

The mappings $(x, y, z) \mapsto x$ and $(x, y, z) \mapsto y$ of $X \times Y \times Z$ into X and Y are continuous (3.2.8), therefore the mapping $(x, y, z) \mapsto (x, y)$ of $X \times Y \times Z$ into $X \times Y$ is continuous (3.2.7). Similarly, the mapping $(x, y, z) \mapsto z$ of $X \times Y \times Z$ into Z is continuous (3.2.8), therefore the mapping $(x, y, z) \mapsto ((x, y), z)$ of $X \times Y \times Z$ into $(X \times Y) \times Z$ is continuous (3.2.7).

Similarly, one proves successively the continuity of the following mappings: $((x, y), z) \mapsto (x, y)$, $((x, y), z) \mapsto x$, $((x, y), z) \mapsto y$, $((x, y), z) \mapsto z$, $((x, y), z) \mapsto (x, y, z)$.

3.2.10. On account of 3.2.9, one identifies the topological spaces $(X \times Y) \times Z$ and $X \times Y \times Z$. This reduces step by step the study of finite products of topological spaces to the study of a product of two spaces.

For $1 \leq p < n$, one identifies $\mathbf{R}^n$ with $\mathbf{R}^p \times \mathbf{R}^{n-p}$, etc.

3.2.11. Theorem. *Let* X, Y *be topological spaces,* y_0 *a fixed point of* Y, A *the subspace* $X \times \{y_0\}$ *of* $X \times Y$. *The mapping* $x \mapsto (x, y_0)$ *of* X *onto* A *is a homeomorphism.*

This mapping is obviously bijective. It is continuous from X into X × Y by 3.2.7, therefore it is continuous from X into A

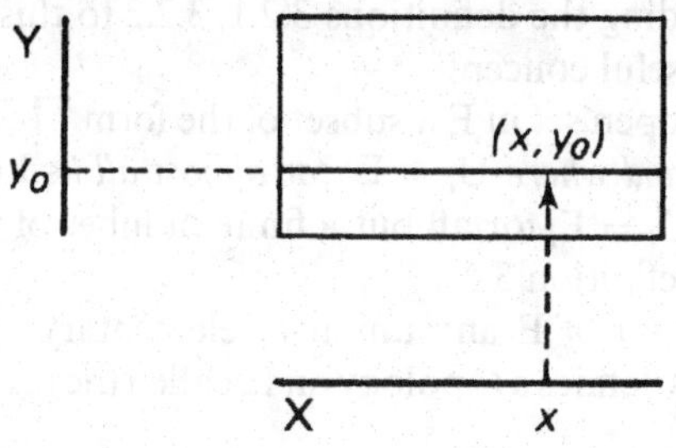

by 3.1.15. The inverse mapping is the composite of the canonical injection of A into X × Y, which is continuous (3.1.16), and of the canonical projection of X × Y onto X, which is also continuous (3.2.8).

3.2.12. For example, one can identify **R** with the subspace **R** × {0} of $\mathbf{R}^2$, etc.

3.2.13. Theorem. *Let* X *be a separated topological space,* Δ *the diagonal of* X × X *(that is, the set of all* (x, x)*, where x runs over* X*). Then* Δ *is closed in* X × X.

Let us show that (X × X) − Δ is open in X × X, that is, is a neighborhood of each of its points. Let $(x, y) \in X \times X$. If $(x, y) \notin \Delta$ then $x \neq y$. Since X is separated, there exist disjoint neighborhoods V, W of x, y. Then V × W is a neighborhood of (x, y) in X × X (3.2.4), and V × W is disjoint from Δ, that is to say contained in (X × X) − Δ. Thus (X × X) − Δ is a neighborhood of (x, y).

3.2.14. Corollary. *Let* E *be a topological space,* F *a separated topological space, f and g continuous mappings of* E *into* F. *The set of* $x \in E$ *such that* $f(x) = g(x)$ *is closed in* E.

For, the mapping $x \mapsto h(x) = (f(x), g(x))$ of E into F × F is continuous (3.2.7). Let Δ be the diagonal of F × F, which is closed in F × F (3.2.13). The set studied in the corollary is nothing more than $h^{-1}(\Delta)$, therefore is closed (2.4.4).

3.2.15. Corollary. *Let* E, F, *f*, *g be as in* 3.2.14. *If f and g are equal on a dense subset of* E, *then* $f = g$.

For, the set of 3.2.14 is here both dense and closed, therefore equal to E.

3.3. Infinite Products of Topological Spaces

3.3.1. Let $(E_i)_{i \in I}$ be a family of topological spaces. Let $E = \prod_{i \in I} E_i$. There is an obvious way of extending the definitions 3.2.1, 3.2.2 to this situation, but this does not lead to a useful concept.

One calls elementary open set in E a subset of the form $\prod_{i \in I} U_i$, where U_i is an open subset of E_i, *and where* $U_i = E_i$ *for almost all* $i \in I$ (which, in this context, will mean that $U_i = E_i$ for all but a finite number of indices). For I finite, one recovers the definition 3.2.1.

One again calls *open set of* E any union of elementary open sets. One verifies as in 3.2.1 that this defines a topology on E, called the product topology of the topologies of the E_i.

3.3.2. Most of the arguments of 3.2 may be extended with minimal complications. We state the results:

(a) Let $x = (x_i)_{i \in I} \in E$, where $x_i \in E_i$ for all $i \in I$. The sets of the form $\prod_{i \in I} V_i$, where V_i is a neighborhood of x_i in E_i and where $V_i = E_i$ for almost all i, constitute a fundamental system of neighborhoods of x in E.

(b) If every E_i is separated, then E is separated.

(c) Let X be a set equipped with a filter base $\mathscr{B}$, f a mapping of X into E (thus of the form $x \mapsto (f_i(x))_{i \in I}$, where f_i is a mapping of X into E_i), and $l = (l_i)_{i \in I} \in E$. The following conditions are equivalent:

(i) f tends to l along $\mathscr{B}$;
(ii) for every $i \in I$, f_i tends to l_i along $\mathscr{B}$.

(d) Let T be a topological space, f a mapping of T into E (thus of the form $t \mapsto (f_i(t))_{i \in I}$, where f_i is a mapping of T into E_i). The following conditions are equivalent: (i) f is continuous; (ii) every f_i is continuous.

(e) The canonical projections of E onto the E_i are continuous.

(f) If I is the union of disjoint subsets I_λ, where λ runs over a set Λ, then the topological space $\prod_{i \in I} E_i$ may be identified with the topological space $\prod_{\lambda \in \Lambda} (\prod_{i \in I_\lambda} E_i)$ ('associativity of the topological product').

3.4. Quotient Spaces

3.4.1. Theorem. *Let* E *be a topological space,* R *an equivalence relation on* E, F *the quotient set* E/R, π *the canonical mapping of* E *onto* F. *Let* $\mathscr{O}$ *be the set of subsets* A *of* F *such that* $\pi^{-1}(A)$ *is open in* E. *Then* $\mathscr{O}$ *satisfies the axioms* (i), (ii), (iii) *of* 1.2.1.

This is immediate.

3.4.2. Thus, $\mathcal{O}$ is the set of open sets of a topology on F, called the *quotient topology of the topology of* E *by* R. Equipped with this topology, F is called the *quotient space of* E *by* R.

The mapping π of E onto F is *continuous* (for, if A is open in F, then $\pi^{-1}(A)$ is open in E).

3.4.3. Example. The set **T** is defined to be the quotient of **R** by the equivalence relation $x - y \in \mathbf{Z}$. By 3.4.2, **T** is equipped with a topology that makes it a quotient space of **R**.

► **3.4.4. Theorem.** *Let* F = E/R *be the quotient space of a topological space* E *by an equivalence relation* R, π *the canonical mapping of* E *onto* F, Y *a topological space, and* f *a mapping of* F *into* Y. *The following conditions are equivalent*:

(i) f *is continuous*;
(ii) *the mapping* $f \circ \pi$ *of* E *into* Y *is continuous.*

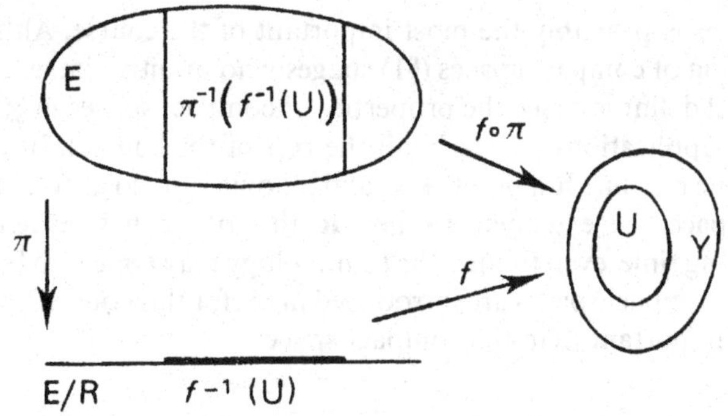

Suppose f is continuous. Since π is continuous (3.4.2), $f \circ \pi$ is continuous. Suppose $f \circ \pi$ is continuous. Let U be an open subset of Y. Then

$$\pi^{-1}(f^{-1}(U)) = (f \circ \pi)^{-1}(U)$$

is open in E (2.4.4), therefore $f^{-1}(U)$ is open in F (3.4.2). Thus f is continuous (2.4.4).

3.4.5. Example. Denote by **U** the set of complex numbers of absolute value 1. One knows that the mapping $x \mapsto g(x) = \exp(2\pi i x)$ of **R** into **U** is surjective, and that $g(x) = g(x') \Leftrightarrow x - x' \in \mathbf{Z}$. Thus, if p denotes the canonical mapping of **R** onto **T**, then g defines, by passage to the quotient, a bijection f of **T** onto **U** such that $f \circ p = g$. Since g is continuous, f is continuous (3.4.4). We shall see later on (4.2.16) that f is a homeomorphism. Let us show that **T** *is separated*. Let x, y be distinct points of **T**. Then $f(x) \neq f(y)$. Since **U** is separated, there exist disjoint open neighborhoods V, W of $f(x)$, $f(y)$ in **U**. Then $f^{-1}(V)$, $f^{-1}(W)$ are disjoint open neighborhoods of x, y in **T**.

CHAPTER IV

Compact Spaces

This chapter is probably the most important of the course. Although the definition of compact spaces (§1) suggests no intuitive image, it is a very fruitful definition (see the properties of compact spaces in §§2 and 3, and the applications in nearly all the rest of the course). In §4, we adjoin to the real line **R** a point $+\infty$ and a point $-\infty$ so as to obtain a compact space, the 'extended real line' $\overline{\mathbf{R}}$; the student has made use of this for a long time, even though the terminology may appear to be new.

Locally compact spaces are introduced in §5; for this course, they are much less important than the compact spaces.

4.1. Definition of Compact Spaces

4.1.1. Theorem. *Let* E *be a topological space. The following conditions are equivalent*:

(i) *If a family of open subsets of* E *covers* E, *one can extract from it a finite subfamily that again covers* E.
(ii) *If a family of closed subsets of* E *has empty intersection, one can extract from it a finite subfamily whose intersection is again empty.*

This is immediate by passage to complements.

4.1.2. Definition. A separated space that satisfies the equivalent conditions of 4.1.1 is called a *compact space.*

Let E be a topological space. A subset A of E such that the topological space A (3.1.2) is compact is of course called a *compact subset* of E.

4.1.3. Theorem. *Let* E *be a topological space,* A *a separated subspace of* E. *The following conditions are equivalent*:

(i) A *is compact*;
(ii) *if a family of open subsets of* E *covers* A, *one can extract from it a finite subfamily that again covers* A.

Suppose A is compact. Let $(U_i)_{i \in I}$ be a family of open subsets of E covering A. The $U_i \cap A$ are open in A (3.1.2) and cover A, therefore there exists a finite subset J of I such that the family $(U_i \cap A)_{i \in J}$ covers A. *A fortiori*, the family $(U_i)_{i \in J}$ covers A.

Suppose conditions (ii) is satisfied. Let $(V_i)_{i \in I}$ be a family of open sets of A covering A. For every $i \in I$, there exists an open set W_i of E such that $V_i = W_i \cap A$ (3.1.2). Then $(W_i)_{i \in I}$ covers A, thus there exists a finite subset J of I such that $(W_i)_{i \in J}$ covers A. Therefore $(V_i)_{i \in J}$ covers A.

4.1.4. Theorem (Borel–Lebesgue). *Let $a, b \in \mathbf{R}$ with $a \leq b$. Then the interval $[a, b]$ is compact.*

It is clear that $[a, b]$ is separated.

Let $(U_i)_{i \in I}$ be a family of open subsets of **R** covering $[a, b]$. Let A be the set of $x \in [a, b]$ such that $[a, x]$ is covered by a finite number of the sets U_i.

The set A is nonempty because $a \in A$. It is contained in $[a, b]$, therefore it is bounded above. Let m be its supremum. Then $a \leq m \leq b$.

There exists $j \in I$ such that $m \in U_j$. Since U_j is open in **R**, there exists $\varepsilon > 0$ such that $[m - \varepsilon, m + \varepsilon] \subset U_j$. Since m is the supremum of A, there exists $x \in A$ such that $m - \varepsilon < x \leq m$. Then $[a, x]$ is covered by a finite number of the U_i, and $[x, m + \varepsilon] \subset U_j$, therefore $[a, m + \varepsilon]$ is covered by a finite number of the U_i. If $m < b$ then, after reducing ε if necessary so that $m + \varepsilon \in [a, b]$, one sees that $m + \varepsilon \in A$, which contradicts the definition of supremum. Therefore $m = b$, and $[a, b]$ is covered by a finite number of the U_i. It then suffices to apply 4.1.3.

4.2. Properties of Compact Spaces

We are going to show: (1) that compact spaces have useful properties; (2) that there exist interesting examples of compact spaces (besides those of 4.1.4). The logical order of the proofs unfortunately obliges us to mix the two objectives.

► **4.2.1. Theorem.** *Let* X *be a set equipped with a filter base* $\mathscr{B}$, E *a compact space, f a mapping of* X *into* E. *Then f admits at least one adherence value along* $\mathscr{B}$.

Consider the family of subsets $\overline{f(B)}$ of E, where B runs over $\mathscr{B}$. These are closed sets. Let $A = \bigcap_{B \in \mathscr{B}} \overline{f(B)}$. If $A = \varnothing$, there exist $B_1, \ldots, B_n \in \mathscr{B}$ such that $\overline{f(B_1)} \cap \cdots \cap \overline{f(B_n)} = \varnothing$ (because E is compact). Now, there exists $B_0 \in \mathscr{B}$ such that $B_0 \subset B_1 \cap \cdots \cap B_n$, whence $f(B_0) \subset f(B_1) \cap \cdots \cap f(B_n)$ and consequently $f(B_1) \cap \cdots \cap f(B_n) \neq \varnothing$. This contradiction proves that $A \neq \varnothing$. In view of 2.6.6, this proves the theorem.

4.2.2. Corollary. *In a compact space, every sequence of points admits at least one adherence value.*

4.2.3. Theorem. *Let* X, $\mathscr{B}$, E, f *be as in* 4.2.1. *Let* A *be the set (nonempty) of adherence values of* f *along* $\mathscr{B}$. *Let* U *be an open subset of* E *containing* A. *There exists* $B \in \mathscr{B}$ *such that* $f(B) \subset U$ *(and even* $\overline{f(B)} \subset U$*).*

One has $(E - U) \cap A = \varnothing$, thus

$$(E - U) \cap \bigcap_{B \in \mathscr{B}} \overline{f(B)} = \varnothing.$$

Since $E - U$ and the $\overline{f(B)}$ are closed we infer, by the compactness of E, that there exist $B_1, \ldots, B_n \in \mathscr{B}$ such that

$$(E - U) \cap \overline{f(B_1)} \cap \cdots \cap \overline{f(B_n)} = \varnothing.$$

Next, there exists $B_0 \in \mathscr{B}$ such that $B_0 \subset B_1 \cap \cdots \cap B_n$. Then

$$(E - U) \cap \overline{f(B_0)} = \varnothing,$$

that is, $\overline{f(B_0)} \subset U$.

► **4.2.4. Corollary.** *Let* X, $\mathscr{B}$, E, f *be as in* 4.2.1. *If* f *admits only one adherence value* l *along* $\mathscr{B}$, *then* f *tends to* l *along* $\mathscr{B}$.

With the preceding notations, we have $A = \{l\}$, and U can be taken to be any open neighborhood of l.

4.2.5. Corollary. *In a compact space, if a sequence of points has only one adherence value* l, *then it tends to* l.

► **4.2.6. Theorem.** *Let* E *be a compact space,* F *a closed subspace of* E. *Then* F *is compact.*

Since E is separated, F is separated. Let $(F_i)_{i \in I}$ be a family of closed subsets of F with empty intersection. Since F is closed in E, the F_i are closed in E (3.1.5). Since E is compact, there exists a finite subfamily $(F_i)_{i \in J}$ with empty intersection.

4.2.7. The converse of 4.2.6 is true. Better yet:

► **Theorem.** *Let* E *be a separated space,* F *a compact subspace of* E. *Then* F *is closed in* E.

We are first going to prove the following:

(*) Let $x_0 \in E - F$. There exist disjoint open sets U, V of E such that $x_0 \in U$ and $F \subset V$.

Since E is separated, for every $y \in F$ there exist open neighborhoods U_y, V_y of x_0, y in E that are disjoint. The V_y, as y runs over F, cover F. Since F is compact, there exist (4.1.3) points $y_1, \ldots, y_n$ of F such that $F \subset V_{y_1} \cup \cdots \cup V_{y_n}$. Let $U = U_{y_1} \cap \cdots \cap U_{y_n}$, which is an open neighborhood of x_0 in E. Set $V = V_{y_1} \cup \cdots \cup V_{y_n}$, which is an open subset of E containing F. Then U and V are disjoint, and we have proved (*).

Since $U \subset E - F$, it follows in particular from (*) that $E - F$ is a neighborhood of x_0 in E. This being true for every $x_0 \in E - F$, $E - F$ is open in E, therefore F is closed in E.

4.2.8. Corollary. *In* **R**, *the compact subsets are the closed bounded subsets.*

Let A be a compact subset of **R**. Then A is closed in **R** (4.2.7). On the other hand, it is clear that $A \subset \bigcup_{x \in A} (x - 1, x + 1)$; by 4.1.3, A is covered by a finite number of intervals $(x_i - 1, x_i + 1)$, hence is bounded.

Let B be a closed, bounded subset of **R**. There exists an interval $[a, b]$ such that $B \subset [a, b]$. Then $[a, b]$ is compact (4.1.4), and B is closed in $[a, b]$ (3.1.5) hence is compact (4.2.6).

4.2.9. Theorem. *Let* E *be a separated space.*

(i) *If* A, B *are compact subsets of* E, *then* $A \cup B$ *is compact.*
(ii) *If* $(A_i)_{i \in I}$ *is a nonempty family of compact subsets of* E, *then* $\bigcap_{i \in I} A_i$ *is compact.*

Let $(U_i)_{i \in I}$ be a covering of $A \cup B$ by open subsets of E. There exist finite subsets J_1, J_2 of I such that $(U_i)_{i \in J_1}$ covers A and $(U_i)_{i \in J_2}$ covers B. Then $(U_i)_{i \in J_1 \cup J_2}$ covers $A \cup B$, which proves that $A \cup B$ is compact (4.1.3).

The A_i are closed in E (4.2.7), therefore $\bigcap_{i \in I} A_i$ is closed in E, hence in each A_i (3.1.5). Since the A_i are compact, $\bigcap_{i \in I} A_i$ is compact (4.2.6).

4.2.10. However, an infinite union of compact subsets is not in general compact. For example, the intervals $[-1, 1]$, $[-2, 2]$, $[-3, 3], \ldots$ of **R** are compact, but their union, which is **R**, is not compact (4.2.8).

4.2.11. Theorem.

(i) *Let* E *be a separated space,* A *and* B *disjoint compact subsets of* E. *There exist disjoint open sets* U, V *of* E *such that* $A \subset U$ *and* $B \subset V$.
(ii) *In a compact space, every point has a fundamental system of compact neighborhoods.*

(i) For every $x \in A$, there exist disjoint open subsets W_x, W'_x of E such that $x \in W_x$, $B \subset W'_x$ (cf. the assertion (*) in the proof of 4.2.7). Since A is compact, there exist $x_1, \ldots, x_p \in A$ such that $A \subset W_{x_1} \cup \cdots \cup W_{x_p}$. Set $U = W_{x_1} \cup \cdots \cup W_{x_p}$ and $V = W'_{x_1} \cap \cdots \cap W'_{x_p}$. Then U and V are open subsets of E, $A \subset U$, $B \subset V$ and $U \cap V = \varnothing$.

(ii) Suppose E is compact. Let $x \in E$ and let Y be an open neighborhood of x in E. Then $\{x\}$ and $E - Y$ are disjoint compact subsets of E. By (i), there exist disjoint open sets U, V of E such that $x \in U$ and $E - Y \subset V$. Then $\overline{U}$ is a compact neighborhood of x. We have $U \subset E - V$, therefore $\overline{U} \subset E - V$ since $E - V$ is closed, whence $\overline{U} \subset Y$.

▶ **4.2.12. Theorem.** *Let* E *be a compact space,* F *a separated space,* f *a continuous mapping of* E *into* F. *Then* $f(E)$ *is compact.*

First, $f(E)$ is separated since F is separated.

Let $(U_i)_{i \in I}$ be a family of open subsets of F covering $f(E)$. Since f is continuous, the $f^{-1}(U_i)$ are open subsets of E (2.4.4). Since the U_i cover $f(E)$, the $f^{-1}(U_i)$ cover E. Since E is compact, there exists a finite subset J of I such that $(f^{-1}(U_i))_{i \in J}$ covers E. Then $(U_i)_{i \in J}$ covers $f(E)$. Therefore $f(E)$ is compact (4.1.3).

4.2.13. Corollary. *Let* E *be a nonempty compact space,* f *a continuous real-valued function on* E. *Then* f *is bounded and attains its infimum and supremum.*

By 4.2.12, $f(E)$ is a compact subset of **R**, hence is a closed, bounded subset of **R** (4.2.8). Since $f(E)$ is bounded, f is bounded. Since, moreover, $f(E)$ is closed, $f(E)$ has a smallest and a largest element (1.5.9). If, for example, $f(x_0)$ is the largest element of $f(E)$, then f attains its supremum at x_0.

4.2.14. Corollary. *Let* E *be a compact space,* f *a continuous real-valued function on* E *with values* >0. *There exists* $\alpha > 0$ *such that* $f(x) \geq \alpha$ *for all* $x \in E$.

Let α be the infimum of f on E. By 4.2.13, there exists $x_0 \in E$ such that $f(x_0) = \alpha$. Therefore $\alpha > 0$. It is clear that $f(x) \geq \alpha$ for all $x \in E$.

4.2.15. Corollary. *Let E be a compact space, F a separated space, f a continuous bijective mapping of E onto F. Then f^{-1} is continuous (in other words, f is a homeomorphism of E onto F).*

Let $g = f^{-1}$. If A is a closed subset of E, then A is compact (4.2.6), therefore $f(A)$ is compact (4.2.12), therefore $f(A)$ is closed in F (4.2.7), in other words $g^{-1}(A)$ is closed in F. This proves that g is continuous (2.4.4).

4.2.16. Example. Let p be the canonical mapping of **R** onto **T**. It is continuous (3.4.2). Since **T** is separated (3.4.5) and $[0, 1]$ is compact (4.1.4), $p([0, 1])$ is compact (4.2.12). But $p([0, 1]) = \mathbf{T}$. Thus *the space* **T** *is compact.*

In 3.4.5 we defined a continuous bijection f of **T** onto **U**. Now, **T** is compact and **U** is separated. Therefore f is a homeomorphism (4.2.15). Thus, the spaces **T** and **U** are homeomorphic.

Since $\mathbf{U}^2$ may be identified with the surface of the space commonly called a 'torus', one says that $\mathbf{T}^2$ is the 2-dimensional torus, and more generally that $\mathbf{T}^n$ is the *n-dimensional torus*. In particular, **T** is called the 1-dimensional torus.

▶ **4.2.17. Theorem.** *The product of a finite number of compact spaces is compact.*

It suffices to show that if X and Y are compact, then X × Y is compact. First, X × Y is separated (3.2.5).

Let $(U_i)_{i \in I}$ be an open covering of X × Y. For every $m = (x, y) \in X \times Y$, choose an $i(m) \in I$ such that $m \in U_{i(m)}$. By 3.2.4, there exist an open neighborhood V_m of x in X and an open neighborhood W_m of y in Y such that $V_m \times W_m \subset U_{i(m)}$. Set $P_m = V_m \times W_m$.

Provisionally, fix $x_0 \in X$. The subset $\{x_0\} \times Y$ of X × Y is homeomorphic to Y (3.2.11) hence is compact. The subsets P_m, where m runs over $\{x_0\} \times Y$, are open in X × Y and cover $\{x_0\} \times Y$, therefore $\{x_0\} \times Y$ is contained in a finite union $P_{m_1} \cup \cdots \cup P_{m_n}$ (4.1.3). The intersection $V_{m_1} \cap \cdots \cap V_{m_n}$ is

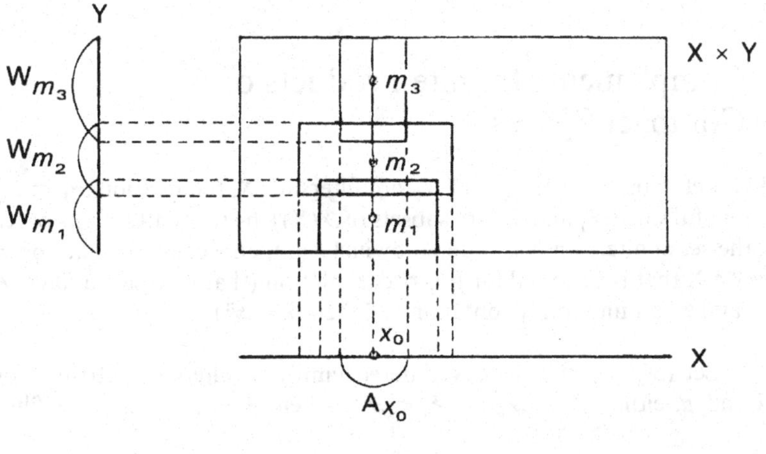

an open neighborhood A_{x_0} of x_0 in X. If $(x, y) \in A_{x_0} \times Y$, there exists a k such that $(x_0, y) \in P_{m_k} = V_{m_k} \times W_{m_k}$, whence $y \in W_{m_k}$ and

$$(x, y) \in A_{x_0} \times W_{m_k} \subset V_{m_k} \times W_{m_k} = P_{m_k}.$$

Thus $A_{x_0} \times Y$ is covered by a finite number of the sets U_i.

If now x_0 runs over X, the A_{x_0} form an open covering of X, from which one can extract a finite covering

$$X = A_{x_1} \cup \cdots \cup A_{x_p}.$$

Each $A_{x_q} \times Y$ is covered by a finite number of the sets U_i, therefore $X \times Y$ is covered by a finite number of the sets U_i.

► **4.2.18. Corollary.** *In $\mathbf{R}^n$, the compact sets are the closed bounded sets.*

(A subset of $\mathbf{R}^n$ is said to be bounded if its n canonical projections onto $\mathbf{R}$ are bounded.)

Let A be a compact subset of $\mathbf{R}^n$. Then A is closed in $\mathbf{R}^n$ (4.2.7). Its canonical projections onto $\mathbf{R}$ are compact (3.2.8 and 4.2.12), hence bounded (4.2.8), thus A is bounded.

Let B be a closed, bounded subset of $\mathbf{R}^n$. Since B is bounded, one has

$$B \subset [a_1, b_1] \times [a_2, b_2] \times \cdots \times [a_n, b_n] = C.$$

Now, C is compact (4.1.4 and 4.2.17), and B is closed in C (3.1.5) hence is compact (4.2.6).

4.2.19. Examples. The space $\mathbf{R}^n$ is not compact.

The sphere S_n (2.5.7) is bounded in $\mathbf{R}^{n+1}$, and closed in $\mathbf{R}^{n+1}$ (1.1.12), hence is compact (4.2.18).

With the notations of 2.5.7, the space $S_n - \{a\}$ is homeomorphic to $\mathbf{R}^n$, hence is not compact.

4.3. Complement: Infinite Products of Compact Spaces

* **4.3.1.** Let X be a set. If $\mathscr{F}_1$ and $\mathscr{F}_2$ are filters on X, the relation $\mathscr{F}_1 \subset \mathscr{F}_2$ is meaningful since $\mathscr{F}_1$ and $\mathscr{F}_2$ are subsets of $\mathscr{P}(X)$ (the set of all subsets of X). Thus, the set of filters on X is ordered by inclusion. One calls *ultrafilter* on X a filter on X that is maximal for this order relation (that is to say, a filter $\mathscr{F}$ such that if $\mathscr{G}$ is a filter on X containing $\mathscr{F}$, then $\mathscr{G} = \mathscr{F}$).

* **4.3.2.** Let $(\mathscr{F}_\lambda)_{\lambda \in \Lambda}$ be a totally ordered family of filters on X (thus, for any λ and μ, either $\mathscr{F}_\lambda \subset \mathscr{F}_\mu$ or $\mathscr{F}_\mu \subset \mathscr{F}_\lambda$). Then $\mathscr{F} = \bigcup_{\lambda \in \Lambda} \mathscr{F}_\lambda$ is a filter

on X. For, let $Y \in \mathscr{F}$ and let $Z \subset X$ be such that $Z \supset Y$; then $Y \in \mathscr{F}_\lambda$ for some λ, therefore $Z \in \mathscr{F}_\lambda$, therefore $Z \in \mathscr{F}$. On the other hand, let $Y_1, Y_2 \in \mathscr{F}$; there exist $\lambda, \mu \in \Lambda$ such that $Y_1 \in \mathscr{F}_\lambda$, $Y_2 \in \mathscr{F}_\mu$; if, for example, $\mathscr{F}_\lambda \subset \mathscr{F}_\mu$, then $Y_1 \in \mathscr{F}_\mu$, therefore $Y_1 \cap Y_2 \in \mathscr{F}_\mu \subset \mathscr{F}$. This proves our assertion.

It then follows from Zorn's theorem (see, for example, Bourbaki, *Theory of Sets*, Ch. III, §2, Cor. 1 of Th. 2) that *every filter on* X *is contained in an ultrafilter.*

* **4.3.3. Theorem.** *Let $\mathscr{F}$ be a filter on* X. *The following conditions are equivalent*:

(i) *$\mathscr{F}$ is an ultrafilter;*
(ii) *for every subset* Y *of* X, *either $Y \in \mathscr{F}$ or $X - Y \in \mathscr{F}$.*

Suppose that $\mathscr{F}$ is not an ultrafilter. Let $\mathscr{F}'$ be a filter on X strictly containing $\mathscr{F}$. There exists $Y \in \mathscr{F}'$ such that $Y \notin \mathscr{F}$. Then $X - Y \notin \mathscr{F}'$ (because $Y \cap (X - Y) = \varnothing$) and *a fortiori* $X - Y \notin \mathscr{F}$.

Suppose there exists a subset Y of X such that $Y \notin \mathscr{F}$ and $X - Y \notin \mathscr{F}$. Let $\mathscr{G}$ be the set of all subsets of X containing a set of the form $F \cap Y$ with $F \in \mathscr{F}$. Let us show that $\mathscr{G}$ is a filter. For every $F \in \mathscr{F}$, one has $F \not\subset X - Y$ (otherwise, $X - Y \in \mathscr{F}$), therefore $F \cap Y \neq \varnothing$; thus, every element of $\mathscr{G}$ is nonempty. It is clear that every subset of X containing an element of $\mathscr{G}$ belongs to $\mathscr{G}$. Finally, let G_1, G_2 be elements of $\mathscr{G}$; then $G_1 \supset F_1 \cap Y$, $G_2 \supset F_2 \cap Y$ with $F_1, F_2 \in \mathscr{F}$, therefore $G_1 \cap G_2 \supset (F_1 \cap F_2) \cap Y$ and $F_1 \cap F_2 \in \mathscr{F}$; therefore $G_1 \cap G_2 \in \mathscr{G}$. We have thus shown that $\mathscr{G}$ is indeed a filter. It is clear that $\mathscr{G} \supset \mathscr{F}$ and that $Y \in \mathscr{G}$, therefore $\mathscr{G} \neq \mathscr{F}$ and $\mathscr{F}$ is not an ultrafilter.

* **4.3.4. Theorem.** *Let* X *and* X′ *be sets, f a mapping of* X *into* X′, *$\mathscr{F}$ an ultrafilter on* X. *Let $\mathscr{F}'$ be the set of all subsets of* X′ *that contain a set of the form $f(F)$, where $F \in \mathscr{F}$. Then $\mathscr{F}'$ is an ultrafilter on* X′.

It is clear that $\mathscr{F}'$ is a filter on X′. Let $Y' \subset X'$. Set $Y = f^{-1}(Y')$. Then $Y \in \mathscr{F}$ or $X - Y \in \mathscr{F}$ (4.3.3). If $Y \in \mathscr{F}$, then $Y' \in \mathscr{F}'$ because $f(Y) \subset Y'$. If $X - Y \in \mathscr{F}$, then $X' - Y' \in \mathscr{F}'$ because $f(X - Y) \subset X' - Y'$. Therefore $\mathscr{F}'$ is an ultrafilter (4.3.3).

* **4.3.5. Theorem.** *Let* E *be a separated topological space. The following conditions are equivalent*:

(i) E *is compact;*
(ii) *if $\mathscr{U}$ is an ultrafilter on a set* X, *and if f is a mapping of* X *into* E, *then f has a limit along $\mathscr{U}$.*

(a) Suppose E is compact. Let X, $\mathscr{U}$, f be as in (ii).

Let x_0 be an adherence value of f along $\mathscr{U}$ (4.2.1). The sets of the form $f(U) \cap V$, where $U \in \mathscr{U}$ and V is a neighborhood of x_0 in E, are nonempty

(2.6.1). They constitute a filter base $\mathscr{B}$ on E, because if $U_1, U_2 \in \mathscr{U}$ and V_1, V_2 are neighborhoods of x_0, then

$$f(U_1 \cap U_2) \cap (V_1 \cap V_2) \subset (f(U_1) \cap V_1) \cap (f(U_2) \cap V_2).$$

Let $\mathscr{F}$ be the set of subsets of E that contain an element of $\mathscr{B}$. Then $\mathscr{F}$ is a filter on E (2.1.1).

Let $\mathscr{G}$ be the set of subsets of E that contain a set of the form $f(U)$, where $U \in \mathscr{U}$. Then $\mathscr{G}$ is an ultrafilter (4.3.4). It is clear that $\mathscr{G} \subset \mathscr{F}$. Therefore $\mathscr{G} = \mathscr{F}$. Now let V be a neighborhood of x_0. Then $V \in \mathscr{F}$, therefore $V \in \mathscr{G}$, therefore there exists $U \in \mathscr{U}$ such that $f(U) \subset V$. Thus f tends to x_0 along $\mathscr{U}$.

(b) Suppose condition (ii) is satisfied. Let $(F_i)_{i \in I}$ be a family of closed subsets of E. Suppose that for every finite subset J of I, the set $F_J = \bigcap_{i \in J} F_i$ is nonempty. We are to show that $\bigcap_{i \in I} F_i \neq \varnothing$. Now, the F_J form a filter base. Let $\mathscr{U}$ be an ultrafilter on E containing all the F_J (4.3.2). Let us apply condition (ii) to the identity mapping of E into E: we see that there exists an $x_0 \in E$ such that every neighborhood of x_0 contains an element of $\mathscr{U}$. Fix $i \in I$. Let V be a neighborhood of x_0. Let $U \in \mathscr{U}$ be such that $U \subset V$. Then $F_i \cap U \neq \varnothing$ since $F_i \in \mathscr{U}$ and $U \in \mathscr{U}$. Therefore $F_i \cap V \neq \varnothing$. This being true for all V, we have $x_0 \in \overline{F}_i = F_i$. This being true for all i, we see that $x_0 \in \bigcap_{i \in I} F_i$.

* **4.3.6. Theorem.** *Let $(E_i)_{i \in I}$ be a family of compact spaces, and let $E = \prod_{i \in I} E_i$. Then E is compact.*

Let X be a set, $\mathscr{U}$ an ultrafilter on X, f a mapping of X into E, hence of the form $x \mapsto (f_i(x))_{i \in I}$, where f_i is a mapping of X into E_i. By 4.3.5, f_i has a limit $l_i \in E_i$ along $\mathscr{U}$. Let $l = (l_i)_{i \in I}$. By 3.3.2(c), f tends to l along $\mathscr{U}$. Therefore E is compact (4.3.5).

4.4. The Extended Real Line

4.4.1. Let $\overline{\mathbf{R}}$ be the set obtained by adjoining to **R** two elements, denoted $+\infty$ and $-\infty$, not belonging to **R**. If $x, y \in \overline{\mathbf{R}}$, the relation $x \leq y$ is defined in the following way:

(1) if $x, y \in \mathbf{R}$, then $x \leq y$ has the usual meaning;
(2) for all $x \in \mathbf{R}$, one sets $x < +\infty$ and $-\infty < x$;
(3) one sets $-\infty < +\infty$.

It is easily verified that one obtains in this way a total order on $\overline{\mathbf{R}}$. The ordered set $\overline{\mathbf{R}}$ is called the *extended real line*.

4.4.2. Consider the mapping f of $[-\pi/2, \pi/2]$ into $\overline{\mathbf{R}}$ defined as follows:

$$f(x) = \tan x \quad \text{if} \quad -\frac{\pi}{2} < x < \frac{\pi}{2},$$

$$f\left(\frac{\pi}{2}\right) = +\infty, \quad f\left(-\frac{\pi}{2}\right) = -\infty.$$

Then f is bijective and increasing, as is f^{-1}, and is therefore an isomorphism of ordered sets. Every property of the ordered set $[-\pi/2, \pi/2]$ is therefore also true in the ordered set $\overline{\mathbf{R}}$. In particular, *every nonempty subset of* $\overline{\mathbf{R}}$ *admits a supremum and an infimum.*

4.4.3. Topology of $\overline{\mathbf{R}}$. Let us call open subset of $\overline{\mathbf{R}}$ the image under f of an open subset of $[-\pi/2, \pi/2]$. One thus defines a topology on $\overline{\mathbf{R}}$, and f is a homeomorphism of $[-\pi/2, \pi/2]$ onto $\overline{\mathbf{R}}$. It follows that $\overline{\mathbf{R}}$ *is compact* and that, in $\overline{\mathbf{R}}$, *every nonempty closed subset has a smallest and a largest element.*

Since the restriction of f to $(-\pi/2, \pi/2)$ is a homeomorphism of $(-\pi/2, \pi/2)$ onto $\mathbf{R}$ (2.5.5), the topology of $\overline{\mathbf{R}}$ induces on $\mathbf{R}$ the usual topology.

The intervals $[b, \pi/2]$, where $-\pi/2 < b < \pi/2$, form a fundamental system of neighborhoods of $\pi/2$ in $[-\pi/2, \pi/2]$. Therefore the intervals $[a, +\infty]$, where $a \in \mathbf{R}$, form a fundamental system of neighborhoods of $+\infty$ in $\overline{\mathbf{R}}$. Similarly, the intervals $[-\infty, a]$, where $a \in \mathbf{R}$, form a fundamental system of neighborhoods of $-\infty$ in $\overline{\mathbf{R}}$. It follows that if $(x_1, x_2, \ldots)$ is a sequence of real numbers, to say that $x_n \to +\infty$ in $\overline{\mathbf{R}}$ means that $x_n \to +\infty$ in the usual sense.

4.4.4. Theorem (Passage to the Limit in Inequalities). *Let* X *be a set equipped with a filter base* $\mathscr{B}$, f *and* f' *mappings of* X *into* $\overline{\mathbf{R}}$, *admitting limits* l, l' *along* $\mathscr{B}$. *Suppose that* $f(x) \leq f'(x)$ *for all* $x \in \mathrm{X}$. *Then* $l \leq l'$.

Suppose $l > l'$. Let $a \in \mathbf{R}$ be such that $l > a > l'$. Then $(a, +\infty]$ and $[-\infty, a)$ are neighborhoods of l and l' in $\overline{\mathbf{R}}$. Therefore there exist $\mathrm{B} \in \mathscr{B}$ and $\mathrm{B}' \in \mathscr{B}$ such that

$$x \in \mathrm{B} \Rightarrow f(x) > a, \qquad x \in \mathrm{B}' \Rightarrow f'(x) < a.$$

If $x \in \mathrm{B} \cap \mathrm{B}'$, one sees that $f(x) > f'(x)$, which is absurd.

4.5. Locally Compact Spaces

4.5.1. Theorem. *Let* E *be a topological space. The following conditions are equivalent:*

(i) *every point of* E *admits a compact neighborhood;*
(ii) *every point of* E *admits a fundamental system of compact neighborhoods.*

Obviously (ii) $\Rightarrow$ (i). Let $x \in$ E and let V be a compact neighborhood of x. By 4.2.11(ii), x admits in V a fundamental system (V_i) of compact neighborhoods. One verifies easily that (V_i) is a fundamental system of neighborhoods of x in E (cf. 3.1.6).

4.5.2. Definition. A topological space is said to be *locally compact* if it is separated and satisfies the equivalent conditions of 4.5.1.

4.5.3. Examples. (a) Every compact space is locally compact.

(b) $\mathbf{R}^n$ is locally compact (without being compact). For, $\mathbf{R}^n$ is separated, and every point of $\mathbf{R}^n$ admits as neighborhood a closed ball, which is compact (4.2.18).

(c) Let us show that $\mathbf{Q}$ is not locally compact. Suppose that the point 0 of $\mathbf{Q}$ possesses in $\mathbf{Q}$ a compact neighborhood V. There exists a neighborhood W of 0 in $\mathbf{R}$ such that $V = W \cap \mathbf{Q}$ (3.1.6). Then there exists $\alpha > 0$ such that $(-\alpha, \alpha) \subset W$, whence $(-\alpha, \alpha) \cap \mathbf{Q} \subset V$. Moreover, since V is compact, V is closed in $\mathbf{R}$ (4.2.7). Now, every real number in $(-\alpha, \alpha)$ is adherent to $(-\alpha, \alpha) \cap \mathbf{Q}$, whence $(-\alpha, \alpha) \subset V$, which is absurd since $V \subset \mathbf{Q}$.

4.5.4. Theorem. *Let* X *be a locally compact space,* Y *an open or closed subset of* X. *Then the space* Y *is locally compact.*

First, Y is separated. On the other hand, let $y \in Y$. There exists a compact neighborhood V of y in X. Then $V \cap Y$ is a neighborhood of y in Y (3.1.6). If Y is closed in X, then $V \cap Y$ is closed in V (3.1.4), hence is compact (4.2.6). If Y is open in X, one can suppose $V \subset Y$ (4.5.1(ii)) and then $V \cap Y = V$.

4.5.5. Theorem. *Let* $X_1, X_2, \ldots, X_n$ *be locally compact spaces and let* $X = X_1 \times \cdots \times X_n$. *Then* X *is locally compact.*

First, X is separated. On the other hand, let $x = (x_1, \ldots, x_n) \in X$. For every i, there exists a compact neighborhood V_i of x_i in X_i. Then $V_1 \times \cdots \times V_n$ is a neighborhood of x in X (3.2.4) and is compact (4.2.17).

* **4.5.6. Remark.** Let E be a compact space, ω a point of E. By 4.5.4, the space $X = E - \{\omega\}$ is locally compact.

* **4.5.7.** Let us show, under the conditions of 4.5.6, how one can reconstruct the topology of E starting from that of X. We are going to prove that the open sets of E are: (1) the open sets of X; (2) the sets of the form $(X - C) \cup \{\omega\}$, where C is a compact subset of X.

First, an open set of X is open in E (3.1.3). Next, if C is a compact subset of X, then C is closed in E (4.2.7), therefore $E - C$ is open in E; now,

$$E - C = (X - C) \cup \{\omega\}.$$

Finally, let U be an open set in E. If $\omega \notin U$ then $U \subset X$, and U is open in X (3.1.3); if $\omega \in U$, then

$$U = (X - C) \cup \{\omega\},$$

where C is the complement of U in E; this complement is closed in E, hence is compact (4.2.6).

* **4.5.8.** The interest of Remark 4.5.6 is that it yields *all* of the locally compact spaces, as we shall now see.

Let X be a locally compact space, X′ a set obtained by adjoining to X a point ω not belonging to X. The construction we shall give is inspired by 4.5.7. Let us say that a subset U of X′ is open in the following two cases: (α) U is an open set of X; (β) U is of the form $(X - C) \cup \{\omega\}$, where C is a compact subset of X. Let us show that the axioms (i), (ii), (iii) of 1.2.1 are satisfied. This is clear for (i). Let $(U_i)_{i \in I}$ be a family of open sets of X′, and U the union of the U_i. Then $I = J \cup K$, where: (1) for $i \in J$, U_i is open in X; (2) for $i \in K$, $U_i = (X - C_i) \cup \{\omega\}$ with C_i compact in X. If $K = \varnothing$ then U is an open subset of X. Suppose $K \neq \varnothing$. Then $\omega \in U$ and

$$X' - U = \bigcap_{i \in I} (X' - U_i) = \left(\bigcap_{i \in K} C_i\right) \cap \left(\bigcap_{i \in J} (X - U_i)\right).$$

Now, $\bigcap_{i \in K} C_i$ is compact in X (4.2.9(ii)) and $\bigcap_{i \in J} (X - U_i)$ is closed in X, therefore $X' - U$ is a compact subset C of X (4.2.6), and $U = (X - C) \cup \{\omega\}$. We have thus verified axiom (ii). Now let U_1, U_2 be open sets of X′ and let us show that $U_1 \cap U_2$ is an open set of X′. This is clear if U_1, U_2 are open in X. If U_1 is open in X and $U_2 = (X - C) \cup \{\omega\}$ with C compact in X, then $U_1 \cap U_2 = U_1 \cap (X - C)$, and $X - C$ is open in X (4.2.7), therefore $U_1 \cap U_2$ is open in X. If $U_1 = (X - C_1) \cup \{\omega\}$ and $U_2 = (X - C_2) \cup \{\omega\}$ with C_1, C_2 compact in X, then

$$U_1 \cap U_2 = (X - (C_1 \cup C_2)) \cup \{\omega\},$$

and $C_1 \cup C_2$ is compact in X (4.2.9(i)).

We have thus defined a topology on X′. The subset X of X′ is open in X′. The intersections with X of the open sets of X′ are the open sets of X. In other words, *the topology induced on* X *by that of* X′ *is the given topology on* X.

Let us show that X′ is separated. Let x, y be distinct points of X′, and let us show that x and y admit disjoint neighborhoods in X′. This is clear if $x, y \in X$. Suppose $x = \omega$ and $y \in X$. Let W be a compact neighborhood of y in X. This is also a neighborhood of y in X′ (because X is open in X′). Set

$$V = (X - W) \cup \{\omega\}.$$

This is an open neighborhood of ω in X′, and $V \cap W = \varnothing$.

Let us show that X′ *is compact*. Let $(U_i)_{i\in I}$ be an open covering of X′. There exists $i_0 \in I$ such that $\omega \in U_{i_0}$. Then

$$U_{i_0} = (X - C) \cup \{\omega\},$$

where C is compact. The U_i cover C, therefore there exists a finite subset J of I such that $(U_i)_{i\in J}$ covers C (4.1.3). Then

$$X' = U_{i_0} \cup \left(\bigcup_{i\in J} U_i\right).$$

* **4.5.9.** One often says that ω is the *point at infinity* of X′, and that X′ results from X by *the adjunction of a point at infinity*. One also says that X′ is the *Alexandroff compactification of* X.

* **4.5.10.** Let X be a locally compact space, E_1 and E_2 compact spaces such that X is a subspace of E_1 and E_2, and such that $E_i - X$ reduces to a point ω_i (for $i = 1$ and 2). Then the unique bijection f of E_1 onto E_2 that reduces to the identity on X and transforms ω_1 into ω_2 is a *homeomorphism* of E_1 onto E_2; for, by 4.5.7, f transforms the open sets of E_1 into the open sets of E_2.

This proves, in a certain sense, the *uniqueness* of the Alexandroff compactification.

* **4.5.11. Example.** We defined in 2.5.7 a homeomorphism of $\mathbf{R}^n$ onto the complement $S_n - \{a\}$ of $\{a\}$ in the sphere S_n. Now, $\mathbf{R}^n$ is locally compact and S_n is compact. In view of 4.5.10, *the Alexandroff compactification of* $\mathbf{R}^n$ *may be identified with* S_n. For example, the Alexandroff compactification of $\mathbf{R}$ may be identified with the circle S_1, hence with $\mathbf{U}$ (hence with the 1-dimensional torus $\mathbf{T}$, by 4.2.16).

CHAPTER V

Metric Spaces

This chapter, in which we reconsider metric spaces in greater detail, is heterogeneous.

In §1, we introduce some concepts that are quite geometrical: diameter of a set, distance between two sets.

In §2, we note that some of the earlier definitions take on a much more intuitive aspect in metric spaces. For example, a point adherent to a set A is nothing more than the limit of a sequence of points of A.

The student already knows what is meant by uniform continuity for a real-valued function of a real variable. This concept is studied in the setting of arbitrary metric spaces in §3, and the somewhat analogous concept of equicontinuity in §4.

The very important concept of complete metric space is studied in §§5, 6, 7. Among the numerous useful theorems concerning such spaces, we cite Baire's theorem (5.5.12).

5.1. Continuity of the Metric

5.1.1. Theorem. *Let* E *be a metric space,* d *its metric. The mapping* $(x, y) \mapsto d(x, y)$ *of* $E \times E$ *into* $\mathbf{R}$ *is continuous.*

Let $(x_0, y_0) \in E \times E$ and let $\varepsilon > 0$. Let V, W be the closed balls with centers x_0, y_0 and radius $\varepsilon/2$. Then $V \times W$ is a neighborhood of (x_0, y_0)

in $E \times E$. If $(x, y) \in V \times W$, then

$$d(x, y) \leq d(x, x_0) + d(x_0, y_0) + d(y_0, y)$$

$$\leq \frac{\varepsilon}{2} + d(x_0, y_0) + \frac{\varepsilon}{2} = d(x_0, y_0) + \varepsilon,$$

$$d(x_0, y_0) \leq d(x_0, x) + d(x, y) + d(y, y_0)$$

$$\leq \frac{\varepsilon}{2} + d(x, y) + \frac{\varepsilon}{2} = d(x, y) + \varepsilon,$$

therefore $|d(x, y) - d(x_0, y_0)| \leq \varepsilon$. This proves the continuity of d at (x_0, y_0).

5.1.2. Definition. Let E be a metric space, A a nonempty subset of E. One calls *diameter* of A the supremum, finite or infinite, of the set of numbers $d(x, y)$, where x and y run over A.

5.1.3. Theorem. *The sets* A *and* $\bar{A}$ *have the same diameter.*

Let $D \subset \mathbf{R}$ (resp. $D' \subset \mathbf{R}$) be the set of $d(x, y)$ where x, y run over A (resp. $\bar{A}$). Then $D \subset D'$. Since every point of $\bar{A} \times \bar{A}$ is adherent to $A \times A$ (3.2.4), we have $D' \subset \bar{D}$ (2.4.4(iv) and 5.1.1). Therefore $\bar{D} = \bar{D'}$. If D is bounded, the diameter of A (resp. $\bar{A}$) is the largest element of $\bar{D}$ (resp. $\bar{D'}$) (1.5.9). If D is unbounded, then D and D′ both have supremum $+\infty$.

5.1.4. Definition. Let E be a metric space, A and B nonempty subsets of E. One calls *distcnce from* A *to* B the infimum of the numbers $d(x, y)$, where x runs over A and y runs over B. It is denoted $d(A, B)$. One has $d(A, B) = d(B, A)$. We remark that if $d(A, B) = 0$, A and B need not be equal and may even be disjoint.

If $z \in E$ and $A \subset E$, we set $d(z, A) = d(\{z\}, A) = \inf_{x \in A} d(z, x)$.

5.1.5. Theorem. *Let* E *be a metric space,* A *and* B *nonempty subsets of* E. *Then* $d(A, B) = d(\bar{A}, \bar{B})$.

The proof is analogous to that of 5.1.3.

5.1.6. Theorem. *Let* E *be a metric space,* A *a nonempty subset of* E, *and* $x, y \in E$. *Then* $|d(x, A) - d(y, A)| \leq d(x, y)$.

Let $\varepsilon > 0$. There exists $z \in A$ such that $d(x, z) \leq d(x, A) + \varepsilon$. Then

$$d(y, A) \leq d(y, z) \leq d(y, x) + d(x, z) \leq d(x, y) + d(x, A) + \varepsilon.$$

Thus $d(y, A) - d(x, A) \leq d(x, y) + \varepsilon$. Interchanging the roles of x and y, one sees that $d(x, A) - d(y, A) \leq d(x, y) + \varepsilon$. Therefore $|d(x, A) - d(y, A)| \leq d(x, y) + \varepsilon$. This being true for every $\varepsilon > 0$, one obtains $|d(x, A) - d(y, A)| \leq d(x, y)$.

5.2. The Use of Sequences of Points in Metric Spaces

▶ **5.2.1. Theorem.** *Let* E *be a metric space,* $A \subset E$, $x \in E$. *The following conditions are equivalent:*

(i) $x \in \overline{A}$;
(ii) *there exists a sequence* $(x_1, x_2, \ldots)$ *of points of* A *that tends to* x.

If condition (ii) is satisfied, every neighborhood of x contains an x_n, therefore intersects A; consequently $x \in \overline{A}$.

If $x \in \overline{A}$ then, for every integer $n \geq 1$, there exists a point x_n of A that belongs to the closed ball with center x and radius $1/n$. Then x_n tends to x by 2.2.2.

▶ **5.2.2. Theorem.** *Let* E *be a metric space,* $(x_1, x_2, \ldots)$ *a sequence of points of* E, *and* $x \in E$. *The following conditions are equivalent:*

(i) x *is an adherence value of* (x_n);
(ii) *there exists a subsequence* $(x_{n_1}, x_{n_2}, \ldots)$, *where* $n_1 < n_2 < \cdots$, *that tends to* x.

Suppose condition (ii) is satisfied. Then x is an adherence value of $(x_{n_1}, x_{n_2}, \ldots)$ and *a fortiori* of $(x_1, x_2, \ldots)$.

Suppose condition (i) is satisfied. There exists n_1 such that $d(x_{n_1}, x) \leq 1$. Then there exists $n_2 > n_1$ such that $d(x_{n_2}, x) \leq \frac{1}{2}$. Then there exists $n_3 > n_2$ such that $d(x_{n_3}, x) \leq \frac{1}{3}$, etc. The sequence $(x_{n_1}, x_{n_2}, \ldots)$ then tends to x.

5.2.3. Theorem. *Let* X, Y *be metric spaces,* A *a subset of* X, f *a mapping of* A *into* Y, $a \in \overline{A}$, *and* $y \in Y$. *The following conditions are equivalent:*

(i) y *is an adherence value of* f *as* x *tends to* a *while remaining in* A (2.6.3);
(ii) *there exists a sequence* (x_n) *of points of* A *such that* $x_n \to a$ *and* $f(x_n) \to y$.

The proof is analogous to that of 5.2.2.

5.2.4. Theorem. *Let X, Y be metric spaces, f a mapping of X into Y, and $x \in X$. The following conditions are equivalent*:

(i) *f is continuous at x*;
(ii) *for every sequence (x_n) of points of X tending to x, the sequence $(f(x_n))$ tends to $f(x)$.*

(i) ⇒ (ii). This is true for any topological spaces (2.3.3).

Not (i) ⇒ not (ii). Suppose that f is not continuous at x. There exists $\varepsilon > 0$ such that, for every $\eta > 0$, there is a $y \in X$ with $d(x, y) \leq \eta$ yet $d(f(x), f(y)) \geq \varepsilon$. Successively take $\eta = 1, \frac{1}{2}, \frac{1}{3}, \ldots$. One obtains points $y_1, y_2, y_3, \ldots$ of X such that $d(x, y_n) \leq 1/n$ and $d(f(x), f(y_n)) \geq \varepsilon$. Then $y_n \to x$ but $f(y_n)$ does not tend to $f(x)$.

► **5.2.5. Theorem.** *Let X be a metric space. The following conditions are equivalent*:

(i) *X is compact*;
(ii) *every sequence of points in X admits at least one adherence value.*

(i) ⇒ (ii). This is true for every compact space (4.2.2).

(ii) ⇒ (i). Suppose condition (ii) is satisfied. Let $(U_i)_{i \in I}$ be an open covering of X, and let us show that X can be covered by a finite number of the U_i. We denote by $B(x, \rho)$ the open ball with center x and radius ρ.

(a) Let us show the existence of an $\alpha > 0$ such that every ball $B(x, \alpha)$ is contained in some U_i.

Suppose that no such α exists. Then, for $n = 1, 2, \ldots$, there exists an $x_n \in X$ such that $B(x_n, 1/n)$ is not contained in any U_i. Let x be an adherence value of $(x_1, x_2, \ldots)$. Then $x \in U_i$ for some i, and so $B(x, 1/N) \subset U_i$ for some N. Next, there exists $n \geq 2N$ such that $x_n \in B(x, 1/2N)$. Then

$$B\left(x_n, \frac{1}{n}\right) \subset B\left(x_n, \frac{1}{2N}\right) \subset B\left(x, \frac{1}{2N} + \frac{1}{2N}\right) \subset U_i,$$

which is absurd.

(b) It now suffices to prove that X can be covered by a finite number of balls $B(x, \alpha)$. Let $x_1 \in X$. If $B(x_1, \alpha) = X$, the proof is over. Otherwise, let $x_2 \in X - B(x_1, \alpha)$. If $B(x_1, \alpha) \cup B(x_2, \alpha) = X$, the proof is over. Otherwise, let $x_3 \in X - (B(x_1, \alpha) \cup B(x_2, \alpha))$; etc. If the process stops, the theorem is established. Otherwise, there exists a sequence $(x_1, x_2, \ldots)$ of points of X such that

$$x_n \notin B(x_1, \alpha) \cup \cdots \cup B(x_{n-1}, \alpha)$$

for every n. The mutual distances of the x_i are $\geq \alpha$. Let $x \in X$ be an adherence value of $(x_1, x_2, \ldots)$. There exists an n such that $x_n \in B(x, \alpha/2)$. Next, there exists an $n' > n$ such that $x_{n'} \in B(x, \alpha/2)$. Then $d(x_n, x_{n'}) < \alpha$, which is absurd.

5.2.6. Corollary. *Let* X *be a metric space,* A *a subset of* X. *The following conditions are equivalent*:

(i) $\overline{A}$ *is compact*;
(ii) *from every sequence of points of* A, *one can extract a subsequence that has a limit in* X.

Suppose $\overline{A}$ is compact. Let (x_n) be a sequence of points of A, hence of $\overline{A}$. It has an adherence value $x \in \overline{A}$. By 5.2.2, some subsequence tends to x.

Suppose condition (ii) is satisfied. Let $(y_1, y_2, \ldots)$ be a sequence of points of $\overline{A}$. Let $x_i \in A$ be such that $d(y_i, x_i) \le 1/i$. Some subsequence $(x_{n_1}, x_{n_2}, \ldots)$ tends to a point $x \in X$. Then $x \in \overline{A}$ (5.2.1). On the other hand,

$$d(y_{n_i}, x) \le d(y_{n_i}, x_{n_i}) + d(x_{n_i}, x) \le \frac{1}{n_i} + d(x_{n_i}, x) \to 0,$$

therefore y_{n_i} tends to x. Thus, the sequence $(y_1, y_2, \ldots)$ has an adherence value in $\overline{A}$. Consequently, $\overline{A}$ is compact (5.2.5).

5.3. Uniformly Continuous Functions

5.3.1. Definition. Let X and Y be metric spaces, f a mapping of X into Y. We say that f is *uniformly continuous* if, for every $\varepsilon > 0$, there exists an $\eta > 0$ such that

$$x_1, x_2 \in X \text{ and } d(x_1, x_2) \le \eta \Rightarrow d(f(x_1), f(x_2)) \le \varepsilon.$$

5.3.2. It is clear that a uniformly continuous mapping of X into Y is continuous at every point, therefore is continuous. But the converse is not true. For example, the mapping $x \mapsto x^2$ of **R** into **R** is continuous, but it is not uniformly continuous. For, suppose it were. Taking $\varepsilon = 1$ in Definition 5.3.1, there would exist $\eta > 0$ such that

$$x_1, x_2 \in \mathbf{R} \text{ and } |x_1 - x_2| \le \eta \Rightarrow |x_1^2 - x_2^2| \le 1.$$

Now, let $x_1 = 1/\eta$, $x_2 = 1/\eta + \eta$. Then $|x_1 - x_2| \le \eta$ and

$$|x_1^2 - x_2^2| = \left|\frac{1}{\eta^2} - \frac{1}{\eta^2} - 2 - \eta^2\right| = 2 + \eta^2 > 1.$$

5.3.3. The example 5.3.2 lends weight to the following theorem:

▶ **Theorem.** *Let* X *and* Y *be metric spaces,* f *a continuous mapping of* X *into* Y. *Assume* X *is compact. Then* f *is uniformly continuous.*

Fix $\varepsilon > 0$. We are to construct an $\eta > 0$ with the property of 5.3.1. For every $x \in X$, there exists an $\eta_x > 0$ such that

$$x' \in X,\ d(x', x) \leq \eta_x \Rightarrow d(f(x'), f(x)) \leq \frac{\varepsilon}{2}.$$

Let B_x be the open ball with center x and radius $\frac{1}{2}\eta_x$. The B_x, as x runs over X, form an open covering of X. Since X is compact, X is covered by a finite number of such balls $B_{x_1}, B_{x_2}, \ldots, B_{x_n}$. Set

$$\eta = \inf(\tfrac{1}{2}\eta_{x_1}, \tfrac{1}{2}\eta_{x_2}, \ldots, \tfrac{1}{2}\eta_{x_n}) > 0.$$

Now let $x', x'' \in X$ be such that $d(x', x'') \leq \eta$. There exists an i such that $x' \in B_{x_i}$, in other words,

(1) $$d(x_i, x') < \tfrac{1}{2}\eta_{x_i}.$$

Then

$$d(x_i, x'') \leq d(x_i, x') + d(x', x'') < \tfrac{1}{2}\eta_{x_i} + \eta \leq \tfrac{1}{2}\eta_{x_i} + \tfrac{1}{2}\eta_{x_i},$$

thus

(2) $$d(x_i, x'') < \eta_{x_i}.$$

The inequalities (1) and (2) imply

$$d(f(x'), f(x_i)) \leq \frac{\varepsilon}{2}, \qquad d(f(x''), f(x_i)) \leq \frac{\varepsilon}{2},$$

whence $d(f(x'), f(x'')) \leq \varepsilon$.

5.4. Equicontinuous Sets of Functions

5.4.1. Let X and Y be metric spaces, f a mapping of X into Y. We recall:

(a) f is continuous at x_0 if, for every $\varepsilon > 0$, there exists an $\eta > 0$ such that $d(x, x_0) \leq \eta \Rightarrow d(f(x), f(x_0)) \leq \varepsilon$;
(b) f is continuous on X if f is continuous at every point of X;
(c) f is uniformly continuous on X if, for every $\varepsilon > 0$, there exists an $\eta > 0$ such that $d(x, x') \leq \eta \Rightarrow d(f(x), f(x')) \leq \varepsilon$.

Now let (f_α) be a *family* of mappings of X into Y.

(a) The family (f_α) is said to be *equicontinuous at* x_0 if, for every $\varepsilon > 0$, there exists an $\eta > 0$ such that

$$d(x, x_0) \leq \eta \Rightarrow d(f_\alpha(x), f_\alpha(x_0)) \leq \varepsilon$$

for all α.

(b) The family (f_α) is said to be *equicontinuous on* X if it is equicontinuous at every point of X.

(c) The family (f_α) is said to be *uniformly equicontinuous on* X if, for every $\varepsilon > 0$, there exists an $\eta > 0$ such that

$$d(x, x') \leq \eta \Rightarrow d(f_\alpha(x), f_\alpha(x')) \leq \varepsilon$$

for all α.

5.4.2. Example. Take X = Y = **R**. Let (f_α) be the family of all differentiable real-valued functions on **R** whose derivative is bounded by 1 in absolute value. Then (f_α) is uniformly equicontinuous. For, let $\varepsilon > 0$; if $x, x' \in \mathbf{R}$ are such that $|x - x'| \leq \varepsilon$, then $|f_\alpha(x) - f_\alpha(x')| \leq \varepsilon$ for all α by the mean value theorem.

5.4.3. Remark. If a family (f_α) is equicontinuous, then each f_α is continuous. However, the converse is not true. For example, take X = Y = **R** and let (f_α) be the family of all linear functions. Each f_α is continuous, but the family (f_α) is not equicontinuous at any point of **R**. For, suppose (f_α) were equicontinuous at x_0. Take $\varepsilon = 1$ in Definition 5.4.1. There would exist an $\eta > 0$ such that

$$|x - x_0| \leq \eta \Rightarrow |f_\alpha(x) - f_\alpha(x_0)| \leq 1$$

for all α. Now, the function $x \mapsto (2/\eta)x$ is linear, and

$$\left|\frac{2}{\eta}(x_0 + \eta) - \frac{2}{\eta}x_0\right| = 2 > 1.$$

5.4.4. Theorem. *Let* X *and* Y *be metric spaces,* (f_α) *an equicontinuous family of mappings of* X *into* Y. *Assume that* X *is compact. Then the family* (f_α) *is uniformly equicontinuous.*

The proof is nearly the same as in 5.3.3.

5.5. Complete Metric Spaces

5.5.1. Definition. Let X be a metric space, $(a_1, a_2, \ldots)$ a sequence of points of X. Recall that the sequence is called a *Cauchy sequence* if

$$d(a_m, a_n) \to 0 \quad \text{as} \quad m \text{ and } n \to \infty,$$

in other words, if

for every $\varepsilon > 0$, there exists an N such that $m, n \geq N \Rightarrow d(a_m, a_n) \leq \varepsilon$.

5.5.2. Theorem. *Let X be a metric space, $(a_1, a_2, \ldots)$ a sequence of points of X. If the sequence has a limit in X, then it is a Cauchy sequence.*

Suppose that (a_n) tends to a. Let $\varepsilon > 0$. There exists a positive integer N such that $n \geq N \Rightarrow d(a_n, a) \leq \varepsilon/2$. Then

$$m, n \geq N \Rightarrow d(a_m, a) \leq \frac{\varepsilon}{2} \text{ and } d(a_n, a) \leq \frac{\varepsilon}{2} \Rightarrow d(a_m, a_n) \leq \varepsilon.$$

5.5.3. It is well known that the converse of 5.5.2 is in general not true (for example in $X = \mathbf{Q}$). We therefore make the following definition:

Definition. A metric space X is said to be *complete* if every Cauchy sequence has a limit in X.

5.5.4. Example. One knows that the metric space **R** is complete.

5.5.5. Theorem. *Let X be a metric space, $(x_1, x_2, \ldots)$ a Cauchy sequence in X, $(x_{n_1}, x_{n_2}, \ldots)$ a subsequence. If the sequence (x_{n_i}) has a limit l, then the sequence (x_n) also tends to l.*

Let $\varepsilon > 0$. There exists an N such that $m, n \geq N \Rightarrow d(x_m, x_n) \leq \varepsilon$. Fix $n \geq N$. For $n_i \geq N$, we have

$$d(x_{n_i}, x_n) \leq \varepsilon. \tag{1}$$

As $i \to \infty$, we have $x_{n_i} \to l$, therefore $d(x_{n_i}, x_n) \to d(l, x_n)$ (5.1.1). By 4.4.4, the inequality (1) implies in the limit that $d(l, x_n) \leq \varepsilon$. This being true for all $n \geq N$, we have $x_n \to l$.

► **5.5.6. Theorem.** *Let X be a complete metric space, Y a closed subspace of X. Then Y is complete.*

Let $(y_1, y_2, \ldots)$ be a Cauchy sequence in Y. It is also a Cauchy sequence in X. Therefore it has a limit a in X. Since $y_i \in Y$ for all i, we have $a \in \overline{Y} = Y$, thus (y_i) has a limit in Y.

5.5.7. The converse of 5.5.6 is true. Better yet:

► **Theorem.** *Let X be a metric space, Y a complete subspace. Then Y is closed in X.*

Let $a \in \overline{Y}$. There exists a sequence $(y_1, y_2, \ldots)$ in Y that tends to a (5.2.1). The sequence (y_i) is a Cauchy sequence (5.5.2). It thus has a limit b in Y since Y is complete. In X, (y_i) tends to a and to b, therefore $a = b \in Y$. Thus $\overline{Y} = Y$.

▶ **5.5.8. Theorem.** *Let* $X_1, \ldots, X_p$ *be complete metric spaces,* $X = X_1 \times \cdots \times X_p$ *the product metric space. Then* X *is complete.*

Let $(y_1, y_2, \ldots)$ be a Cauchy sequence in X. Each point y_i is of the form $(y_{i1}, y_{i2}, \ldots, y_{ip})$, where $y_{i1} \in X_1, \ldots, y_{ip} \in X_p$. We have

$$d(y_{m1}, y_{n1}) \leq d(y_m, y_n) \to 0$$

as $m, n \to \infty$, therefore $(y_{11}, y_{21}, y_{31}, \ldots)$ is a Cauchy sequence in X_1. Since X_1 is complete, this sequence has a limit l_1 in X_1. Similarly, $(y_{12}, y_{22}, y_{32}, \ldots)$ has a limit l_2 in $X_2, \ldots, (y_{1p}, y_{2p}, y_{3p}, \ldots)$ has a limit l_p in X_p. Therefore $(y_1, y_2, \ldots)$ tends in X to the point $l = (l_1, \ldots, l_p)$ (3.2.6).

5.5.9. Examples. By 5.5.4 and 5.5.8, $\mathbf{R}^n$ is complete. By 5.5.6, every closed subset of $\mathbf{R}^n$ is a complete space.

5.5.10. Theorem. *Let* X *be a complete metric space. Let* $(F_1, F_2, \ldots)$ *be a decreasing sequence of nonempty closed subsets of* X *with diameters* $\delta_1, \delta_2, \ldots$. *Assume that* $\delta_i \to 0$ *as* $i \to \infty$. *Then the intersection of the* F_i *consists of exactly one point.*

Let $F = F_1 \cap F_2 \cap F_3 \cap \cdots$. If F is nonempty then its diameter is $\leq \delta_i$ for all i, hence is zero. There are thus two possibilities: either F is empty or it reduces to one point. Let us show that F is nonempty.

Let $a_i \in F_i$. Let us show that $(a_1, a_2, \ldots)$ is a Cauchy sequence. Let $\varepsilon > 0$. There exists an N such that $\delta_N \leq \varepsilon$. If $m, n \geq N$ then $a_m, a_n \in F_N$, therefore $d(a_m, a_n) \leq \varepsilon$.

Since X is complete, the sequence (a_i) tends to a limit a in X. Since $a_i \in F_n$ for $i \geq n$, we have $a \in \overline{F}_n = F_n$. This being true for every n, we have $a \in F$.

5.5.11. Theorem. *Let* X *be a complete metric space,* T *a set equipped with a filter base* $\mathscr{B}$, f *a mapping of* T *into* X. *Assume that for every* $\varepsilon > 0$, *there exists* $B \in \mathscr{B}$ *such that* $f(B)$ *has diameter* $\leq \varepsilon$. *Then* f *has a limit along* $\mathscr{B}$.

Denote by $\delta(A)$ the diameter of a subset A of X. There exists $B_1 \in \mathscr{B}$ such that $\delta(f(B_1)) \leq 1$. Next, there exists $B_2 \in \mathscr{B}$ such that $\delta(f(B_2)) \leq \frac{1}{2}$; replacing B_2 by an element of $\mathscr{B}$ contained in $B_1 \cap B_2$, we can suppose that $B_2 \subset B_1$. Next, there exists $B_3 \in \mathscr{B}$ such that $\delta(f(B_3)) \leq \frac{1}{3}$ and $B_3 \subset B_2$, etc. By 5.1.3, $\delta(\overline{f(B_i)}) \leq 1/i$. By 5.5.10, the intersection of the $\overline{f(B_i)}$ consists of a point l. Let us show that f tends to l along $\mathscr{B}$. Let $\varepsilon > 0$. There exists a positive integer n such that $1/n \leq \varepsilon$. If $x \in B_n$ then $f(x) \in f(B_n)$ and $l \in \overline{f(B_n)}$, therefore

$$d(f(x), l) \leq \delta(\overline{f(B_n)}) \leq \frac{1}{n} \leq \varepsilon.$$

Thus $f(B_n)$ is contained in the closed ball with center l and radius ε.

► **5.5.12. Theorem** (Baire). *Let* X *be a complete metric space,* $U_1, U_2, \ldots$ *a sequence of dense open subsets of* X. *Then* $U_1 \cap U_2 \cap \cdots$ *is dense in* X.

Set $U = U_1 \cap U_2 \cap \cdots$. We are to prove that for every open ball B with radius > 0, $U \cap B$ is nonempty. Denote by $B(x, \rho)$ (resp. $B'(x, \rho)$) the open ball (resp. closed ball) with center x and radius ρ.

The set $U_1 \cap B$ is open and nonempty, therefore it contains a ball $B_1 = B'(x_1, \rho_1)$ such that $0 < \rho_1 \leq 1$. The set $U_2 \cap B(x_1, \rho_1)$ is open and nonempty, therefore it contains a ball $B_2 = B'(x_2, \rho_2)$ such that $0 < \rho_2 \leq \frac{1}{2}$. The set $U_3 \cap B(x_2, \rho_2)$ is open and nonempty, therefore it contains a ball $B_3 = B'(x_3, \rho_3)$ such that $0 < \rho_3 \leq \frac{1}{3}$, etc.

By construction, $B \supset B_1 \supset B_2 \supset B_3 \supset \cdots$. The B_i are closed and their diameter tends to zero. By 5.5.10, there exists a point a that belongs to all of the B_i. Then $a \in B$. Moreover, $a \in B_n \subset U_n$ for all n; therefore $a \in U$.

5.5.13. Theorem. *Let* X *be a metric space,* X' *a dense subspace of* X, Y *a complete metric space,* f' *a uniformly continuous mapping of* X' *into* Y.

(i) *There exists one and only one continuous mapping* f *of* X *into* Y *that extends* f'.
(ii) f *is uniformly continuous.*

The uniqueness of f in (i) follows from 3.2.15. Let us prove the existence of f. For each $x \in X$, choose a sequence (x_n) in X' that tends to x. Let $\varepsilon > 0$. There exists an $\eta > 0$ such that

$$z_1, z_2 \in X' \text{ and } d(z_1, z_2) \leq \eta \Rightarrow d(f'(z_1), f'(z_2)) \leq \varepsilon.$$

Now, (x_n) is a Cauchy sequence, therefore there exists an N such that

$$m, n \geq N \Rightarrow d(x_m, x_n) \leq \eta.$$

Then

$$m, n \geq N \Rightarrow d(f'(x_m), f'(x_n)) \leq \varepsilon.$$

Thus $(f'(x_n))$ is a Cauchy sequence in Y, consequently has a limit in Y which we denote $f(x)$. We have thus defined a mapping f of X into Y. If $x \in X'$ then $f'(x_n)$ tends to $f'(x)$, therefore $f(x) = f'(x)$; in other words, f extends f'.

Let $u, v \in X$ be such that $d(u, v) \leq \eta/2$. Let $(u_1, u_2, \ldots)$ and $(v_1, v_2, \ldots)$ be the chosen sequences in X' tending to u and v. Then $d(u_n, v_n) \to d(u, v)$ (5.1.1), therefore there exists an N such that

$$n \geq N \Rightarrow d(u_n, v_n) \leq \eta \Rightarrow d(f'(u_n), f'(v_n)) \leq \varepsilon.$$

Letting n tend to infinity, one obtains $d(f(u), f(v)) \leq \varepsilon$. Thus

$$u, v \in X \text{ and } d(u, v) \leq \frac{\eta}{2} \Rightarrow d(f(u), f(v)) \leq \varepsilon,$$

which proves that f is uniformly continuous, and *a fortiori* continuous.

5.6. Complete Spaces and Compact Spaces

▶ **5.6.1. Theorem.** *Let X be a metric space. The following conditions are equivalent*:

(i) X *is compact*;
(ii) X *is complete and, for every* $\varepsilon > 0$, *there exists a finite covering of* X *by balls of radius* ε.

Suppose X is compact. Let $(x_1, x_2, \ldots)$ be a Cauchy sequence in X. One can extract a subsequence that has a limit in X (5.2.6). Therefore $(x_1, x_2, \ldots)$ has a limit in X (5.5.5). Thus X is complete. Let $\varepsilon > 0$. The open balls with radius ε form a covering of X; since X is compact, a finite number of such balls suffices for covering X.

Suppose condition (ii) is satisfied. Let $(x_1, x_2, \ldots)$ be a sequence of points of X. Cover X by a finite number of balls of radius $\frac{1}{2}$; one of these balls contains x_i for infinitely many i. One can therefore extract from (x_i) a subsequence of points whose mutual distances are ≤ 1. Let us start anew with $\frac{1}{2}$ replaced by $\frac{1}{4}, \frac{1}{6}, \frac{1}{8}, \ldots$. We obtain an infinity of sequences

$$y_1^1, y_2^1, y_3^1, \ldots$$
$$y_1^2, y_2^2, y_3^2, \ldots$$
$$y_1^3, y_2^3, y_3^3, \ldots$$

each of which is a subsequence of the preceding one, and such that $d(y_i^n, y_j^n) \leq 1/n$ for all i and j. The 'diagonal' sequence $(y_1^1, y_2^2, y_3^3, \ldots)$ is a subsequence of (x_i), and $d(y_m^m, y_n^n) \leq 1/m$ for $m \leq n$. Therefore (y_i^i) is a Cauchy sequence; consequently, it has a limit. Then X is compact by 5.2.6 applied with A = X.

One can obviously replace condition (ii) by the following: X is complete and, for every $\eta > 0$, there exists a finite covering of X by sets of diameter $\leq \eta$.

5.6.2. Theorem. *Let* X *be a complete metric space,* A *a subset of* X. *The following conditions are equivalent*:

(i) $\overline{A}$ *is compact*;
(ii) *for every* $\varepsilon > 0$, *one can cover* A *by a finite number of balls of* X *with radius* ε.

(i) ⇒ (ii). This is obvious.

(ii) ⇒ (i). Suppose condition (ii) is satisfied. Let $\varepsilon > 0$. There exist closed balls $B_1, \ldots, B_n$ in X with radius ε that cover A. Then $\overline{A} \subset B_1 \cup \cdots \cup B_n$. On the other hand, $\overline{A}$ is complete (5.5.6). Therefore $\overline{A}$ is compact (5.6.1).

5.7. The Method of Successive Approximations

5.7.1. Theorem. *Let X be a complete metric space, f a mapping of X into X. Assume that there exists a $\lambda \in [0, 1)$ such that $d(f(x), f(x')) \leq \lambda d(x, x')$ for all $x, x' \in X$.*

(i) *There exists one and only one $a \in X$ such that $f(a) = a$.*
(ii) *For any $x_0 \in X$, the sequence of points*

$$x_1 = f(x_0), x_2 = f(x_1), \ldots$$

tends to a.

Let $a, b \in X$ be such that $a = f(a)$, $b = f(b)$. Then

$$d(a, b) = d(f(a), f(b)) \leq \lambda d(a, b),$$

thus $(1 - \lambda)d(a, b) \leq 0$. Since $1 - \lambda > 0$ we infer that $d(a, b) \leq 0$, whence $d(a, b) = 0$ and $a = b$.

Let $x_0 \in X$. Set $x_1 = f(x_0)$, $x_2 = f(x_1), \ldots$. We are going to prove that (x_n) tends to a limit a and that $f(a) = a$. The theorem will thus be established.

Let us show that $d(x_n, x_{n+1}) \leq \lambda^n d(x_0, x_1)$. This is clear for $n = 0$. If it is true for n, then

$$\begin{aligned} d(x_{n+1}, x_{n+2}) &= d(f(x_n), f(x_{n+1})) \leq \lambda d(x_n, x_{n+1}) \\ &\leq \lambda\lambda^n d(x_0, x_1) = \lambda^{n+1} d(x_0, x_1), \end{aligned}$$

whence our assertion by induction. From this, one deduces that if n and p are integers ≥ 0, then

$$\begin{aligned} d(x_n, x_{n+p}) &\leq d(x_n, x_{n+1}) + d(x_{n+1}, x_{n+2}) + \cdots + d(x_{n+p-1}, x_{n+p}) \\ &\leq (\lambda^n + \lambda^{n+1} + \cdots + \lambda^{n+p-1})\, d(x_0, x_1) \\ &\leq \lambda^n(1 + \lambda + \lambda^2 + \cdots)\, d(x_0, x_1) = \lambda^n \frac{d(x_0, x_1)}{1 - \lambda}. \end{aligned}$$

Since $0 \leq \lambda < 1$, $\lambda^n \to 0$ as $n \to \infty$. We thus see that the sequence (x_n) is a Cauchy sequence, consequently tends to a limit a.

We have $d(x_n, a) \to 0$, therefore $d(f(x_n), f(a)) \to 0$, that is, $d(x_{n+1}, f(a)) \to 0$. Thus, the sequence (x_n) tends also to $f(a)$, whence $f(a) = a$.

5.7.2. Example. Let I be a closed interval of **R**, f a function defined on I, with values in I, such that $\sup_{x \in I} |f'(x)| < 1$. By the mean value theorem, f satisfies the condition of 5.7.1. Consequently, the equation $x = f(x)$ has a unique solution in I which can be obtained, starting with any point x_0 of I, by the 'successive approximations' $x_1 = f(x_0)$, $x_2 = f(x_1), \ldots$.

* **5.7.3. Remark.** Let X be a complete metric space, B the closed ball in X with center x_0 and radius ρ, f a mapping *of* B into X such that $d(f(x), f(x')) \leq \lambda d(x, x')$ for all $x, x' \in B$ (where $0 \leq \lambda < 1$). Assume, moreover, that

$$d(f(x_0), x_0) \leq (1 - \lambda)\rho.$$

Then there exists one and only one $a \in B$ such that $f(a) = a$.

Uniqueness is proved as in 5.7.1. One again forms $x_1 = f(x_0)$, $x_2 = f(x_1), \ldots$, but, for these points to be defined, one must prove that they do not exit B. Indeed, let us suppose that $x_p \in B$ and $d(x_p, x_{p+1}) \leq \lambda^p d(x_0, x_1)$ for $p = 0, 1, \ldots, n$. Then

$$d(x_0, x_{n+1}) \leq (1 + \lambda + \cdots + \lambda^n)d(x_0, x_1) \leq \frac{1}{1-\lambda} d(x_0, x_1) \leq \rho,$$

therefore $x_{n+1} \in B$, and $d(x_{n+1}, x_{n+2}) \leq \lambda^{n+1} d(x_0, x_1)$ may be proved as in 5.7.1.

This established, (x_n) is again a Cauchy sequence and tends to a limit $a \in B$; $f(a) = a$ is proved as in 5.7.1.

CHAPTER VI

Limits of Functions

For real-valued functions of a real variable, the student already knows what it means for a sequence $f_1, f_2, \ldots$ of functions to tend uniformly, or to tend simply, to a function f. In this chapter we study these concepts in the general setting of metric spaces. We obtain in this way certain of the 'infinite-dimensional' spaces alluded to in the Introduction, and, thanks to Ascoli's theorem, the *compact subsets* of these spaces.

6.1. Uniform Convergence

6.1.1. Let X and Y be two sets. The mappings of X into Y form a set which will henceforth be denoted $\mathscr{F}(\mathrm{X}, \mathrm{Y})$.

6.1.2. Let X be a set, Y a metric space. For $f, g \in \mathscr{F}(\mathrm{X}, \mathrm{Y})$, we set

$$d(f, g) = \sup_{x \in \mathrm{X}} d(f(x), g(x)) \in [0, +\infty].$$

Let us show that d is a *metric* (with possibly infinite values) on $\mathscr{F}(\mathrm{X}, \mathrm{Y})$. If $d(f, g) = 0$ then, for every $x \in \mathrm{X}$,

$$d(f(x), g(x)) = 0,$$

therefore $f(x) = g(x)$; thus $f = g$. It is clear that $d(f, g) = d(g, f)$. Finally, if $h \in \mathscr{F}(\mathrm{X}, \mathrm{Y})$ then, for every $x \in \mathrm{X}$,

$$d(f(x), h(x)) \leq d(f(x), g(x)) + d(g(x), h(x)) \leq d(f, g) + d(g, h);$$

this being true for all $x \in \mathrm{X}$, we infer that

$$d(f, h) \leq d(f, g) + d(g, h).$$

This metric is called the *metric of uniform convergence* on $\mathscr{F}(X, Y)$. The corresponding topology is called the *topology of uniform convergence*.

6.1.3. Let $f, f_1, f_2, f_3, \ldots \in \mathscr{F}(X, Y)$. To say that (f_n) tends to f for this topology means that $\sup_{x \in X} d(f(x), f_n(x)) \to 0$, in other words: for every $\varepsilon > 0$ there exists an N such that

$$n \geq N \Rightarrow d(f_n(x), f(x)) \leq \varepsilon \quad \text{for all } x \in X.$$

We then also say that the sequence (f_n) *tends uniformly to* f.

6.1.4. Let Λ be a set equipped with a filter base $\mathscr{B}$. For every $\lambda \in \Lambda$, let $f_\lambda \in \mathscr{F}(X, Y)$. Let $f \in \mathscr{F}(X, Y)$. To say that f_λ tends to f along $\mathscr{B}$ for the topology of uniform convergence means: for every $\varepsilon > 0$, there exists $B \in \mathscr{B}$ such that

$$\lambda \in B \Rightarrow d(f_\lambda(x), f(x)) \leq \varepsilon \quad \text{for all } x \in X.$$

We then also say that f_λ tends to f uniformly along $\mathscr{B}$.

6.1.5. Example. Take $X = Y = \Lambda = \mathbf{R}$. For filter base $\mathscr{B}$ on Λ, take the set of intervals $[a, +\infty)$. For $\lambda \in \mathbf{R}$ and $x \in \mathbf{R}$, set $f_\lambda(x) = e^{-\lambda(x^2+1)}$. Then f_λ tends to 0 uniformly as $\lambda \to +\infty$ (that is, along $\mathscr{B}$). For, let $\varepsilon > 0$. There exists $a \in \mathbf{R}$ such that $\lambda \geq a \Rightarrow e^{-\lambda} \leq \varepsilon$. Then (provided $a \geq 0$):

$$\lambda \geq a \Rightarrow |e^{-\lambda(x^2+1)} - 0| = e^{-\lambda(x^2+1)} \leq e^{-\lambda} \leq \varepsilon \quad \text{for all } x \in \mathbf{R}.$$

6.1.6. Theorem. *Let X be a set, Y a complete metric space. Then the metric space $\mathscr{F}(X, Y)$ is complete.*

Let (f_n) be a Cauchy sequence in $\mathscr{F}(X, Y)$. Let $x \in X$. Then

$$d(f_m(x), f_n(x)) \leq d(f_m, f_n) \to 0 \quad \text{as} \quad m, n \to \infty,$$

thus $(f_n(x))$ is a Cauchy sequence in Y, consequently has a limit in Y which we denote $f(x)$. We have thus defined a mapping f of X into Y.

Let $\varepsilon > 0$. There exists an N such that

$$\begin{aligned} m, n \geq N &\Rightarrow d(f_m, f_n) \leq \varepsilon \\ &\Rightarrow d(f_m(x), f_n(x)) \leq \varepsilon \quad \text{for all } x \in X. \end{aligned}$$

We provisionally fix $x \in X$ and $m \geq N$. As $n \to \infty$, the preceding inequality yields in the limit

$$d(f_m(x), f(x)) \leq \varepsilon.$$

This being true for all $x \in X$, we have $d(f_m, f) \leq \varepsilon$. Thus,

$$m \geq N \Rightarrow d(f_m, f) \leq \varepsilon.$$

In other words, (f_m) tends to f in $\mathscr{F}(X, Y)$.

6.1.7. Thus, to verify that a sequence of mappings of X into Y tends uniformly to a limit, it suffices to verify a 'Cauchy criterion' for the sequence.

6.1.8. Let $u_1, u_2, \ldots \in \mathscr{F}(X, C)$. Let $s \in \mathscr{F}(X, C)$. We say that the series with general term u_n *converges uniformly and has sum s* if the sequence of finite partial sums $u_1 + u_2 + \cdots + u_n$ tends uniformly to s as $n \to \infty$, in other words if, for every $\varepsilon > 0$, there exists an N such that

$$n \geq N \Rightarrow |u_1(x) + u_2(x) + \cdots + u_n(x) - s(x)| \leq \varepsilon \quad \text{for all } x \in X.$$

For the series with general term u_n to converge uniformly, it is necessary and sufficient, by 6.1.6, that the following condition be satisfied: for every $\varepsilon > 0$, there exists an N such that

$$n \geq m \geq N \Rightarrow |u_m(x) + u_{m+1}(x) + \cdots + u_n(x)| \leq \varepsilon \quad \text{for all } x \in X.$$

6.1.9. Let $u_1, u_2, \ldots \in \mathscr{F}(X, C)$. We say that the series with general term u_n *converges normally* if there exists a sequence $\alpha_1, \alpha_2, \ldots$ of numbers ≥ 0 such that $\sum_1^\infty \alpha_n < +\infty$ and such that $|u_n(x)| \leq \alpha_n$ for all $n = 1, 2, \ldots$ and all $x \in X$.

6.1.10. Theorem. *A normally convergent series is uniformly convergent.*

Let $u_1, u_2, \ldots \in \mathscr{F}(X, C)$. Let $\alpha_1, \alpha_2, \ldots$ be numbers ≥ 0 such that $\sum_1^\infty \alpha_n < +\infty$ and $|u_n(x)| \leq \alpha_n$ for all n and x. Let $\varepsilon > 0$. There exists an N such that

$$n \geq m \geq N \Rightarrow \alpha_m + \alpha_{m+1} + \cdots + \alpha_n \leq \varepsilon.$$

Then

$$n \geq m \geq N \Rightarrow |u_m(x) + \cdots + u_n(x)| \leq$$
$$|u_m(x)| + \cdots + |u_n(x)| \leq \alpha_m + \cdots + \alpha_n \leq \varepsilon$$

for all $x \in X$. By 6.1.8, the series with general term u_n is uniformly convergent.

▶ **6.1.11. Theorem.** *Let* X *be a topological space,* Y *a metric space,* Λ *a set equipped with a filter base* $\mathscr{B}$. *For every* $\lambda \in \Lambda$, *let* $f_\lambda \in \mathscr{C}(X, Y)$. *Assume that* f_λ *tends to* $f \in \mathscr{F}(X, Y)$ *uniformly along* $\mathscr{B}$. *Then* $f \in \mathscr{C}(X, Y)$.

Let $x_0 \in X$ and $\varepsilon > 0$. There exists $\lambda \in \Lambda$ such that

$$d(f_\lambda(x), f(x)) \leq \frac{\varepsilon}{3}$$

for all $x \in X$. Next, there exists a neighborhood V of x_0 in X such that

$$x \in V \Rightarrow d(f_\lambda(x), f_\lambda(x_0)) \leq \frac{\varepsilon}{3}.$$

Then, for all $x \in V$, we have

$$d(f(x), f(x_0)) \leq d(f(x), f_\lambda(x)) + d(f_\lambda(x), f_\lambda(x_0)) + d(f_\lambda(x_0), f(x_0))$$

$$\leq \frac{\varepsilon}{3} + \frac{\varepsilon}{3} + \frac{\varepsilon}{3} = \varepsilon,$$

thus f is continuous at x_0.

6.1.12. Corollary. *Let X be a topological space, Y a metric space.*

(i) $\mathscr{C}(X, Y)$ *is closed in* $\mathscr{F}(X, Y)$.
(ii) *If Y is complete, then* $\mathscr{C}(X, Y)$ *is complete.*

The assertion (i) follows from 6.1.11 and 5.2.1. Assertion (ii) follows from (i), 6.1.6 and 5.5.6.

6.1.13. Theorem. *Let T and X be topological spaces, Y a metric space, and* $g \in \mathscr{C}(T \times X, Y)$. *For every* $t \in T$, *set* $f_t(x) = g(t, x)$ *(where* $x \in X$*), so that the* f_t *are mappings of X into Y. Let* $t_0 \in T$. *Assume that X is compact. Then* f_t *tends uniformly to* f_{t_0} *as* t *tends to* t_0.

Let $\varepsilon > 0$. For each $x \in X$, g is continuous at (t_0, x), therefore there exist an open neighborhood V_x of t_0 in T and an open neighborhood W_x of x in X such that

$$(1) \qquad t \in V_x \text{ and } x' \in W_x \Rightarrow d(g(t, x'), g(t_0, x)) \leq \frac{\varepsilon}{2}.$$

As x runs over X, the W_x form an open covering of X, therefore there exist $x_1, \ldots, x_n \in X$ such that $X = W_{x_1} \cup \cdots \cup W_{x_n}$. Set $V = V_{x_1} \cap \cdots \cap V_{x_n}$; this is a neighborhood of t_0 in T.

Let $t \in V$, $z \in X$. There exists an i such that $z \in W_{x_i}$. Moreover, $t \in V_{x_i}$. Therefore $d(g(t, z), g(t_0, x_i)) \leq \varepsilon/2$ by (1). Similarly, (1) implies that $d(g(t_0, z), g(t_0, x_i)) \leq \varepsilon/2$. Then

$$d(g(t, z), g(t_0, z)) \leq \varepsilon.$$

This being true for every $z \in X$, we have $d(f_t, f_{t_0}) \leq \varepsilon$. Thus $t \in V \Rightarrow d(f_t, f_{t_0}) \leq \varepsilon$, so that f_t tends to f_{t_0} uniformly as t tends to t_0.

6.1.14. Theorem (Interchange of Order of Limits). *Let S and T be sets equipped with filter bases* $\mathscr{B}$, $\mathscr{C}$. *Let Y be a complete metric space, g a mapping of* $S \times T$ *into Y. For every* $s \in S$, *set* $f_s(t) = g(s, t)$ *(where* $t \in T$*), so that* $f_s \in \mathscr{F}(T, Y)$. *We make the following assumptions:*

each f_s *has a limit* l_s *along* $\mathscr{C}$;
f_s *tends uniformly along* $\mathscr{B}$ *to an* $f \in \mathscr{F}(T, Y)$.

Then:

(i) *f has a limit l along $\mathscr{C}$*;
(ii) *$s \mapsto l_s$ has a limit l' along $\mathscr{B}$*;
(iii) *$l = l'$*.

Let $\varepsilon > 0$. There exists $\mathrm{B} \in \mathscr{B}$ such that $s \in \mathrm{B} \Rightarrow d(f_s, f) \leq \varepsilon/3$. Fix $s_0 \in \mathrm{B}$. Then

$$d(f_{s_0}(t), f(t)) \leq \frac{\varepsilon}{3} \quad \text{for all } t \in \mathrm{T}. \tag{1}$$

There exists $\mathrm{C} \in \mathscr{C}$ such that $t, t' \in \mathrm{C} \Rightarrow d(f_{s_0}(t), f_{s_0}(t')) \leq \varepsilon/3$. Then if $t, t' \in \mathrm{C}$, we have

$$d(f(t), f(t')) \leq d(f(t), f_{s_0}(t)) + d(f_{s_0}(t), f_{s_0}(t')) + d(f_{s_0}(t'), f(t'))$$

$$\leq \frac{\varepsilon}{3} + \frac{\varepsilon}{3} + \frac{\varepsilon}{3} = \varepsilon.$$

Since Y is complete, we deduce from this (5.5.11) that f has a limit l along $\mathscr{C}$.

Again let $s_0 \in \mathrm{B}$. Along $\mathscr{C}$, the pair

$$(f_{s_0}(t), f(t)) \in \mathrm{Y} \times \mathrm{Y}$$

tends to (l_{s_0}, l). The inequality (1) yields in the limit $d(l_{s_0}, l) \leq \varepsilon/3$. This being true for every $s_0 \in \mathrm{B}$, we conclude that l_s tends to l along $\mathscr{B}$.

* **6.1.15. Remark.** Let X be a topological space, Y a metric space, (f_n) a sequence of mappings of X into Y, f a mapping of X into Y. We say that (f_n) tends to f *uniformly on every compact set* if, for every compact subset C of X, the sequence of restrictions $f_n|\mathrm{C}$ tends uniformly to $f|\mathrm{C}$. There exists a 'topology of compact convergence' such that the preceding concept is precisely the concept of a sequence tending to a limit for this topology; however, we shall not define it.

If (f_n) tends to f uniformly on X, then of course (f_n) tends to f uniformly on every compact set. However, the converse is not true. For example, take $\mathrm{X} = \mathrm{Y} = \mathbf{R}$ and $f_n(x) = e^{-(x-n)^2}$ for $n = 1, 2, \ldots$. Let $a > 0$ and $\mathrm{A} > 0$. Then

$$x \in [-a, a] \text{ and } n \geq \mathrm{A} + a \Rightarrow (x - n)^2 \geq \mathrm{A}^2 \Rightarrow e^{-(x-n)^2} \leq e^{-\mathrm{A}^2};$$

therefore (f_n) tends uniformly to 0 on $[-a, a]$, and this for every a. Therefore (f_n) tends to 0 uniformly on each compact subset of $\mathbf{R}$. However, (f_n) does not tend to 0 uniformly on $\mathbf{R}$, because $f_n(n) = 1$ and so $\sup_{x \in \mathbf{R}} |f_n(x)| \geq 1$.

* **6.1.16.** Nevertheless, we have the following result:

Theorem. *Let* X *be a locally compact space,* Y *a metric space,* (f_n) *a sequence of continuous mappings of* X *into* Y, f *a mapping of* X *into* Y. *Assume that* (f_n) *tends to* f *uniformly on every compact set. Then* f *is continuous.*

Let $x_0 \in X$. There exists a compact neighborhood V of x_0 in X. The sequence $(f_n | V)$ tends uniformly to $f | V$, therefore $f | V$ is continuous (6.1.11). By 2.2.10, f is continuous at x_0. Thus f is continuous.

6.2. Simple Convergence

6.2.1. Let X and Y be sets. Consider the family of sets $(Y_x)_{x \in X}$, where $Y_x = Y$ for all $x \in X$. An element of $\prod_{x \in X} Y_x$ is the giving, for each $x \in X$, of an element of Y; in other words, it is a mapping of X into Y. Thus:

$$\prod_{x \in X} Y_x = \mathscr{F}(X, Y).$$

Now suppose Y is a topological space. Then $\prod_{x \in X} Y_x$, in other words $\mathscr{F}(X, Y)$, carries a product topology (3.3.1), called the *topology of simple convergence* (or the 'topology of pointwise convergence').

6.2.2. Theorem. *Let X be a set, Y a topological space, and $f \in \mathscr{F}(X, Y)$. Let $x_1, \ldots, x_n \in X$ and let W_i be a neighborhood of $f(x_i)$ in Y. Let*

$$V(x_1, \ldots, x_n, W_1, \ldots, W_n)$$

be the set of all $g \in \mathscr{F}(X, Y)$ such that

$$g(x_1) \in W_1, \ g(x_2) \in W_2, \ \ldots, \ g(x_n) \in W_n.$$

Then the $V(x_1, \ldots, x_n, W_1, \ldots, W_n)$ constitute a fundamental system of neighborhoods of f in $\mathscr{F}(X, Y)$ for the topology of simple convergence.

This follows from 3.3.2(a).

6.2.3. Theorem. *If Y is separated, then $\mathscr{F}(X, Y)$ is separated for the topology of simple convergence.*

This follows from 3.3.2(b).

6.2.4. Theorem. *Let Λ be a set equipped with a filter base $\mathscr{B}$. For every $\lambda \in \Lambda$, let $f_\lambda \in \mathscr{F}(X, Y)$. Let $f \in \mathscr{F}(X, Y)$. The following conditions are equivalent:*

(i) *f_λ tends to f along $\mathscr{B}$ for the topology of simple convergence;*
(ii) *for every $x \in X$, $f_\lambda(x)$ tends to $f(x)$ along $\mathscr{B}$.*

This follows from 3.3.2(c).

6.2.5. In particular, if $f, f_1, f_2, f_3, \ldots$ are mappings of X into Y, to say that the sequence (f_n) tends to f for the topology of simple convergence means that, for every $x \in X$, $f_n(x)$ tends to $f(x)$. We then also say that (f_n) *tends simply to* f.

6.2.6. Theorem. *Let* $x_0 \in X$. *The mapping* $f \mapsto f(x_0)$ *of* $\mathscr{F}(X, Y)$ *into* Y is *continuous for the topology of simple convergence.*

This follows from 3.3.2(e).

6.2.7. Let X be a set, Y a metric space. On $\mathscr{F}(X, Y)$, there is the topology $\mathscr{T}_1$ of uniform convergence and the topology $\mathscr{T}_2$ of simple convergence. Then $\mathscr{T}_1$ is *finer than* $\mathscr{T}_2$. By 2.4.7, it suffices to show that the identity mapping of $\mathscr{F}(X, Y)$ equipped with $\mathscr{T}_1$ into $\mathscr{F}(X, Y)$ equipped with $\mathscr{T}_2$ is continuous. For this, it suffices by 6.2.4 to show that, for any fixed x_0 in X, the mapping $f \mapsto f(x_0)$ of $\mathscr{F}(X, Y)$ equipped with $\mathscr{T}_1$ into Y is continuous. Now, this is clear since

$$d(f(x_0), g(x_0)) \leq d(f, g) \quad \text{for } f, g \in \mathscr{F}(X, Y).$$

6.2.8. Let X be a set, Y a metric space. It follows from 6.2.7 that if a sequence (f_n) of elements of $\mathscr{F}(X, Y)$ tends uniformly to an element f of $\mathscr{F}(X, Y)$, then (f_n) tends simply to f. The converse is in general not true (see the example in 6.1.15).

6.2.9. Nevertheless, we have the following result:

▶ **Theorem** (Dini). *Let* X *be a compact space. Let*

$$f, f_1, f_2, \ldots \in \mathscr{C}(X, \mathbf{R}).$$

Assume that $f_1 \leq f_2 \leq f_3 \leq \cdots$ *and that* (f_n) *tends simply to* f. *Then* (f_n) *tends uniformly to* f.

We have $f_n(x) \leq f(x)$ for all $x \in X$. Set $g_n = f - f_n$. The g_n are continuous, tend simply to 0, and

$$g_1 \geq g_2 \geq g_3 \geq \cdots \geq 0.$$

Let $\varepsilon > 0$. Let X_n be the set of $x \in X$ such that $g_n(x) \geq \varepsilon$. Then $X_1 \supset X_2 \supset X_3 \supset \cdots$ and the X_n are closed (cf. 2.4.5). If $x \in \bigcap X_n$ then $g_n(x) \geq \varepsilon$ for all n, which is absurd. Therefore $\bigcap X_n = \varnothing$. Since X is compact, the intersection of a finite number of the X_n is empty. Since the X_n decrease, this intersection is one of the X_n. Thus $X_{n_0} = \varnothing$ for some n_0. Then, for $n \geq n_0$, we have $0 \leq g_n(x) \leq \varepsilon$ for all $x \in X$, thus $|f_n(x) - f(x)| \leq \varepsilon$ for all $x \in X$.

6.2.10. Changing f_n to $-f_n$, we see that Dini's theorem remains valid for decreasing sequences.

6.3. Ascoli's Theorem

► **6.3.1. Theorem** (Ascoli). *Let* X *be a compact metric space,* Y *a complete metric space,* A *an equicontinuous subset of* $\mathscr{C}(X, Y)$. *Assume that for each* $x \in X$ *the set of* $f(x)$, *where* f *runs over* A, *has compact closure in* Y. *Then* A *has compact closure in the metric space* $\mathscr{C}(X, Y)$.

The metric space $\mathscr{C}(X, Y)$ is complete (6.1.12). By 5.6.2, it suffices to prove the following: given any $\varepsilon > 0$, A can be covered by a finite number of balls of radius ε.

By 5.4.4, there exists an $\eta > 0$ such that

$$x, x' \in X, d(x, x') \leq \eta, f \in A \Rightarrow d(f(x), f(x')) \leq \frac{\varepsilon}{4}.$$

We can cover X by a finite number of open balls with centers $x_1, \ldots, x_n$ and radius η. The set of values of the elements of A at $x_1, \ldots, x_n$ has compact closure in Y (4.2.9(i)); we cover it by a finite number of open balls with centers $y_1, \ldots, y_p$ and radius $\varepsilon/4$.

Let Γ be the set of all mappings of $\{1, 2, \ldots, n\}$ into $\{1, 2, \ldots, p\}$; this is a finite set. For each $\gamma \in \Gamma$ let A_γ be the set of $f \in A$ such that

$$d(f(x_1), y_{\gamma(1)}) \leq \frac{\varepsilon}{4}, \ldots, d(f(x_n), y_{\gamma(n)}) \leq \frac{\varepsilon}{4}.$$

By construction, the A_γ cover A. It remains only to show that for any fixed γ, A_γ is contained in some ball of radius ε.

Now, let $f, g \in A_\gamma$. Let $x \in X$. There exists an x_i such that $d(x, x_i) < \eta$. Therefore

$$d(f(x), f(x_i)) \leq \frac{\varepsilon}{4}, \qquad d(g(x), g(x_i)) \leq \frac{\varepsilon}{4}.$$

Moreover,

$$d(f(x_i), y_{\gamma(i)}) \leq \frac{\varepsilon}{4}, \qquad d(g(x_i), y_{\gamma(i)}) \leq \frac{\varepsilon}{4}.$$

Therefore $d(f(x), g(x)) \leq \varepsilon$. This being true for all $x \in X$, we have $d(f, g) \leq \varepsilon$.

6.3.2. Example. Take $X = [0, 1]$, $Y = \mathbf{R}$. Let A be the set of differentiable real-valued functions on $[0, 1]$ such that $|f(x)| \leq 1$ and $|f'(x)| \leq 1$ for all $x \in [0, 1]$. As in 5.4.2, A is equicontinuous. By 6.3.1, A has compact closure in $\mathscr{C}([0, 1], \mathbf{R})$.

In particular (5.2.6), every sequence of functions belonging to A has a uniformly convergent subsequence.

CHAPTER VII

Numerical Functions

This chapter, devoted to real-valued functions, is heterogeneous.

In §§1 and 2 we take up again some familiar concepts, perhaps in a little more general setting.

Let $(u_1, u_2, u_3, \ldots)$ be a sequence of real numbers. The sequence does not always have a limit, but it does have adherence values in $\overline{\mathbf{R}}$; among these, two play an important role: they are called (perhaps inappropriately) the limit superior and the limit inferior of the sequence. A more general definition is presented in §3.

In §4 we define semicontinuous functions, which generalize (for real-valued functions) the continuous functions. In connection with Theorem 7.4.15 (which is a corollary of Baire's theorem) we point out that even if we limited ourselves to continuous functions, the proof would naturally introduce semicontinuous functions.

The student is already familiar with various theorems on the approximation of real-valued functions of a real variable: by ordinary polynomials, or by trigonometric polynomials (cf. the theory of Fourier series). In §5 we give a very general result that encompasses these earlier results. It is applied in §6 (devoted to 'normal' spaces) to the approximation of continuous functions on product spaces.

The mappings of a set X into $\overline{\mathbf{R}}$ are called numerical functions. If the mapping has values in $\mathbf{R}$ we sometimes say, more precisely, finite numerical function.

7.1. Bounds of a Numerical Function

7.1.1. Let f be a numerical function on X. Recall that the *supremum of f on* X, denoted $\sup_{x\in X} f(x)$, is the supremum of the set $f(X)$. This is the element a of $\overline{\mathbf{R}}$ characterized by the following two properties:

(1) $f(x) \leq a$ for all $x \in X$;
(2) for any $b < a$, there exists an $x \in X$ such that $f(x) > b$.

The infimum of f on X, denoted $\inf_{x\in X} f(x)$, is the infimum of the set $f(X)$. One has

$$\inf_{x\in X} f(x) = -\sup_{x\in X}(-f(x)),$$

which reduces the properties of the infimum to the properties of the supremum.

7.1.2. Recall that f is said to be *bounded above* if $\sup_{x\in X} f(x) < +\infty$, *bounded below* if $\inf_{x\in X} f(x) > -\infty$, and *bounded* if it is both bounded above and bounded below. A bounded function is finite, but a finite function is not necessarily bounded.

7.1.3. One calls *oscillation of f over* X the number

$$\sup_{x\in X} f(x) - \inf_{x\in X} f(x)$$

(provided that f is neither constantly $+\infty$ nor constantly $-\infty$, so that the difference is defined).

7.1.4. Theorem. *Let* X *be a set, f a numerical function on* X, $(X_i)_{i\in I}$ *a family of subsets of* X *covering* X. *Then*:

$$\sup_{x\in X} f(x) = \sup_{i\in I}\left(\sup_{x\in X_i} f(x)\right).$$

Set $a = \sup_{x\in X} f(x)$, $a_i = \sup_{x\in X_i} f(x)$. It is clear that $a_i \leq a$ for all $i \in I$. Let $b < a$. There exists $x \in X$ such that $f(x) > b$. Next, there exists $i \in I$ such that $x \in X_i$. Then $b < a_i$. Thus $a = \sup_{i\in I} a_i$.

7.1.5. Corollary. *Let* X, Y *be sets, f a numerical function on* X × Y. *Then*:

$$\sup_{(x,y)\in X\times Y} f(x, y) = \sup_{x\in X}\left(\sup_{y\in Y} f(x, y)\right)$$

$$= \sup_{y\in Y}\left(\sup_{x\in X} f(x, y)\right).$$

The set X × Y is the union, as x runs over X, of the sets $\{x\} \times Y$. The first equality thus follows from 7.1.4.

7.1.6. Theorem. *Let* X *be a set, f and g numerical functions on* X.

(i) $\sup_{x\in X} (f(x) + g(x)) \le \sup_{x\in X} f(x) + \sup_{x\in X} g(x)$.
(ii) *If $f \ge 0$ and $g \ge 0$, then*

$$\sup_{x\in X} (f(x)g(x)) \le \left(\sup_{x\in X} f(x)\right)\left(\sup_{x\in X} g(x)\right).$$

Set $a = \sup_{x\in X} f(x)$, $b = \sup_{x\in X} g(x)$. For every $x \in X$, we have $f(x) \le a$, $g(x) \le b$, therefore $f(x) + g(x) \le a + b$. Consequently $\sup_{x\in X} (f(x)+g(x)) \le a + b$. Assertion (ii) is proved in an analogous manner.

(The statement presumes that the numbers $f(x) + g(x)$, for example, are defined; this would not be the case if, at some point x_0 of X, one had for example $f(x_0) = +\infty$ and $g(x_0) = -\infty$. Also excluded are expressions such as $0 \cdot +\infty$. Here, and in what follows, it is implicitly understood that we are avoiding such indeterminate expressions.)

7.1.7. Theorem. *Let* X *be a set, f a numerical function on* X, *and $k \in \overline{\mathbf{R}}$.*

(i) $\sup_{x\in X} (f(x) + k) = (\sup_{x\in X} f(x)) + k$.
(ii) *If $k \ge 0$, then* $\sup_{x\in X} (kf(x)) = k(\sup_{x\in X} f(x))$.

Set $a = \sup_{x\in X} f(x)$. By 7.1.6,

$$\sup_{x\in X} (f(x) + k) \le a + k.$$

If $k = +\infty$, equality clearly holds; suppose $k < +\infty$. Now let $b < a + k$. We have $b = c + k$ with $c < a$. There exists $x_0 \in X$ with $f(x_0) \ge c$, whence $f(x_0) + k \ge c + k = b$; therefore $\sup_{x\in X} (f(x) + k) \ge b$. This proves (i). One reasons analogously for (ii).

7.1.8. Corollary. *Let* X, Y *be sets, f a numerical function on* X, *and g a numerical function on* Y.

(i) $\sup_{x\in X, y\in Y} (f(x) + g(y)) = \sup_{x\in X} f(x) + \sup_{y\in Y} g(y)$.
(ii) *If $f \ge 0$ and $g \ge 0$, then*

$$\sup_{x\in X, y\in Y} (f(x)g(y)) = \left(\sup_{x\in X} f(x)\right) \cdot \left(\sup_{y\in Y} g(y)\right).$$

Set

$$a = \sup_{x\in X} f(x), \qquad b = \sup_{y\in Y} g(y),$$

$$c = \sup_{x\in X, y\in Y} (f(x) + g(y)).$$

By 7.1.5,

$$c = \sup_{y \in Y} \left[\sup_{x \in X} (f(x) + g(y)) \right].$$

Inside the brackets, $g(y)$ is a constant. By 7.1.7,

$$c = \sup_{y \in Y} (a + g(y)).$$

Then, again by 7.1.7, $c = a + b$. One reasons similarly for (ii).

7.2. Limit of an Increasing Numerical Function

7.2.1. Let X be an ordered set. We say that X is *increasingly filtering* (or 'directed upward') if, for every $x \in X$ and $x' \in X$, there exists an $x'' \in X$ such that $x'' \geq x$ and $x'' \geq x'$. Decreasingly filtering ordered sets are defined in an analogous way.

7.2.2. Examples. (a) A totally ordered set is both increasingly filtering and decreasingly filtering.

(b) Let I be a set, X the set of all finite subsets of I. Order X by inclusion. Then X is both increasingly filtering and decreasingly filtering.

(c) Let $\mathcal{B}$ be a filter base on a set. Order $\mathcal{B}$ by inclusion. Then $\mathcal{B}$ is decreasingly filtering.

7.2.3. Let X be an increasingly filtering ordered set. For every $x \in X$, let B_x be the set of majorants of x in X (that is, the set of elements of X that are $\geq x$). Then, the B_x form a filter base $\mathcal{B}$ on X. For, $x \in B_x$, thus $B_x \neq \emptyset$. On the other hand, if $x, x' \in X$, there exists a majorant x'' of x and x', and one has $B_{x''} \subset B_x \cap B_{x'}$.

When f is a mapping of X into a topological space, the limit of f along $\mathcal{B}$—if it exists—is called *the limit of f along the increasingly filtering set* X and is denoted $\lim_X f$ or $\lim_X f(x)$.

There are analogous definitions for decreasingly filtering sets.

7.2.4. Theorem. *Let* X *be an increasingly filtering ordered set, f an increasing mapping of* X *into* $\overline{\mathbf{R}}$, *and l the supremum of f. Then the limit of f along* X *exists and is equal to l.*

We can suppose $l > -\infty$. Let V be a neighborhood of l in $\overline{\mathbf{R}}$; it contains a neighborhood of the form $[a, b]$, where $a < l \leq b$. There exists an $x \in X$ such that $f(x) \geq a$. Then, for all $y \geq x$ in X, we have

$$f(y) \geq f(x) \geq a,$$

whereas $f(y) \leq l$, therefore $f(y) \in [a, b] \subset V$.

7.2.5. Changing f to $-f$, we see that if f is decreasing and l' is its infimum, then the limit of f along X exists and is equal to l'.

7.2.6. Suppose X is decreasingly filtering. If f is increasing (resp. decreasing), then the limit of f along X exists and is equal to the infimum (resp. supremum) of f. Indeed, for the opposite order on X, X is then increasingly filtering and f is decreasing (resp. increasing).

7.3. Limit Superior and Limit Inferior of a Numerical Function

7.3.1. Definition. Let X be a set equipped with a filter base $\mathscr{B}$, f a mapping of X into $\overline{\mathbf{R}}$, A the set of adherence values of f along $\mathscr{B}$. By 2.6.6 and 4.2.1, A is closed in $\overline{\mathbf{R}}$ and is nonempty, hence admits a smallest and a largest element (4.4.3). These elements are called the *limit inferior of f along* $\mathscr{B}$ and the *limit superior of f along* $\mathscr{B}$. They are denoted $\liminf_{\mathscr{B}} f$ and $\limsup_{\mathscr{B}} f$ (or $\liminf_{\mathscr{B}} f(x)$, $\limsup_{\mathscr{B}} f(x)$).

This definition admits many special cases:

(a) If (u_n) is a sequence of real numbers, one can speak of $\limsup_{n\to\infty} u_n$ and $\liminf_{n\to\infty} u_n$ (these are elements of $\overline{\mathbf{R}}$ and always exist, whereas $\lim_{n\to\infty} u_n$ does not always exist).

(b) If f is a mapping of a topological space X into $\overline{\mathbf{R}}$ and if $a \in$ X, one can speak of $\limsup_{x\to a} f(x)$ and $\liminf_{x\to a} f(x)$.

Etc.

▶ **7.3.2. Theorem.** *Let* X *be a set equipped with a filter base* $\mathscr{B}$, *f a mapping of* X *into* $\overline{\mathbf{R}}$.

(i) $\limsup_{\mathscr{B}} f(x) \geq \liminf_{\mathscr{B}} f(x)$.

(ii) *For f to have a limit along* $\mathscr{B}$, *it is necessary and sufficient that*

$$\limsup_{\mathscr{B}} f(x) = \liminf_{\mathscr{B}} f(x),$$

and the common value is then the limit of f.

(i) This is obvious.

(ii) The space $\overline{\mathbf{R}}$ is compact (4.4.3). Therefore, in order that f have a limit along $\mathscr{B}$, it is necessary and sufficient that the set of adherence values of f along $\mathscr{B}$ reduce to a single point, which is then the limit (2.6.4 and 4.2.4). This implies (ii) at once.

7.3.3. Theorem. *Let* X *be a set equipped with a filter base* $\mathscr{B}$, *f a mapping of* X *into* $\overline{\mathbf{R}}$, *and* $m, n \in \overline{\mathbf{R}}$ *such that*

$$m > \limsup_{\mathscr{B}} f(x) \quad \textit{and} \quad n < \liminf_{\mathscr{B}} f(x).$$

Then there exists $\mathrm{B} \in \mathscr{B}$ *such that*

$$x \in \mathrm{B} \Rightarrow n < f(x) < m.$$

For, (n, m) is an open interval of $\overline{\mathbf{R}}$ that contains the set of all adherence values of f along $\mathscr{B}$, and it suffices to apply 4.2.3.

7.3.4. Theorem. *Let* X *be a set equipped with a filter base* $\mathscr{B}$, *f a mapping of* X *into* $\overline{\mathbf{R}}$. *For every* $\mathrm{B} \in \mathscr{B}$, *let*

$$u_{\mathrm{B}} = \sup_{x \in \mathrm{B}} f(x), \qquad v_{\mathrm{B}} = \inf_{x \in \mathrm{B}} f(x).$$

Then (*cf.* 7.2.2(c) *and* 7.2.3):

$$\limsup_{\mathscr{B}} f(x) = \inf_{\mathrm{B} \in \mathscr{B}} u_{\mathrm{B}} = \lim_{\mathscr{B}} u_{\mathrm{B}},$$

$$\liminf_{\mathscr{B}} f(x) = \sup_{\mathrm{B} \in \mathscr{B}} v_{\mathrm{B}} = \lim_{\mathscr{B}} v_{\mathrm{B}}.$$

Since f can be replaced by $-f$, it suffices to prove the first group of formulas. Set

$$a = \limsup_{\mathscr{B}} f(x), \qquad b = \inf_{\mathrm{B} \in \mathscr{B}} u_{\mathrm{B}}.$$

If $\mathrm{B}, \mathrm{B}' \in \mathscr{B}$ and $\mathrm{B} \supset \mathrm{B}'$, then $u_{\mathrm{B}} \geq u_{\mathrm{B}'}$; by 7.2.4, $\lim_{\mathscr{B}} u_{\mathrm{B}}$ exists and is equal to b. Since a is an adherence value of f along $\mathscr{B}$, we have $a \in \overline{f(\mathrm{B})}$ for all $\mathrm{B} \in \mathscr{B}$ (2.6.6); since u_{B} is the largest element of $\overline{f(\mathrm{B})}$ (1.5.9 and 4.4.3), we have $a \leq u_{\mathrm{B}}$; this being true for every $\mathrm{B} \in \mathscr{B}$, we have $a \leq b$. Suppose $a < b$. Let $\alpha \in (a, b)$. As in 7.3.3, there exists $\mathrm{B} \in \mathscr{B}$ such that $x \in \mathrm{B} \Rightarrow f(x) < \alpha$; then $u_{\mathrm{B}} \leq \alpha$ and *a fortiori* $b \leq \alpha$, which is absurd.

7.3.5. Example. Let (u_n) be a sequence of real numbers. Then:

$$\limsup_{n \to \infty} u_n = \inf_p \left(\sup_{n \geq p} u_n \right) = \lim_{p \to \infty} \left(\sup_{n \geq p} u_n \right),$$

$$\liminf_{n \to \infty} u_n = \sup_p \left(\inf_{n \geq p} u_n \right) = \lim_{p \to \infty} \left(\inf_{n \geq p} u_n \right).$$

► **7.3.6. Theorem.** *Let X be a set equipped with a filter base $\mathscr{B}$, f and g mappings of X into $\overline{\mathbf{R}}$ such that $f(x) \le g(x)$ for all $x \in$ X. Then:*

$$\limsup_{\mathscr{B}} f(x) \le \limsup_{\mathscr{B}} g(x),$$

$$\liminf_{\mathscr{B}} f(x) \le \liminf_{\mathscr{B}} g(x).$$

Let $B \in \mathscr{B}$. Then $f(x) \le \sup_{x \in B} g(x)$ for all $x \in B$, therefore $\sup_{x \in B} f(x) \le \sup_{x \in B} g(x)$. Passing to the limit in this inequality, and taking account of 7.3.4, one obtains $\limsup_{\mathscr{B}} f(x) \le \limsup_{\mathscr{B}} g(x)$. One sees similarly that $\liminf_{\mathscr{B}} f(x) \le \liminf_{\mathscr{B}} g(x)$.

7.3.7. Theorem. *Let X be a set equipped with a filter base $\mathscr{B}$, f and g mappings of X into $\overline{\mathbf{R}}$. Then*

$$\limsup_{\mathscr{B}} (f(x) + g(x)) \le \limsup_{\mathscr{B}} f(x) + \limsup_{\mathscr{B}} g(x),$$

and, if $f \ge 0$, $g \ge 0$, then

$$\limsup_{\mathscr{B}} (f(x)g(x)) \le \left(\limsup_{\mathscr{B}} f(x)\right) \cdot \left(\limsup_{\mathscr{B}} g(x)\right).$$

If one of the functions f, g has a limit along $\mathscr{B}$, these inequalities become equalities.

Let $B \in \mathscr{B}$. Then

$$\sup_{x \in B} (f(x) + g(x)) \le \sup_{x \in B} f(x) + \sup_{x \in B} g(x)$$

by 7.1.6; passing to the limit along the ordered set $\mathscr{B}$, and using 7.3.4, we deduce that

(1) $$\limsup_{\mathscr{B}} (f(x) + g(x)) \le \limsup_{\mathscr{B}} f(x) + \limsup_{\mathscr{B}} g(x).$$

Set $v_B = \inf_{x \in B} f(x)$. For all $x \in B$,

$$v_B + g(x) \le f(x) + g(x),$$

therefore, in view of 7.1.7,

$$v_B + \sup_{x \in B} g(x) \le \sup_{x \in B} (f(x) + g(x));$$

passing to the limit along the ordered set $\mathscr{B}$, we deduce that

(2) $$\liminf_{\mathscr{B}} f(x) + \limsup_{\mathscr{B}} g(x) \le \limsup_{\mathscr{B}} (f(x) + g(x)).$$

If f has a limit along $\mathscr{B}$, then $\limsup_{\mathscr{B}} f(x) = \liminf_{\mathscr{B}} f(x)$. Comparing (1) and (2), we see that

$$\limsup_{\mathscr{B}} (f(x) + g(x)) = \lim_{\mathscr{B}} f(x) + \limsup_{\mathscr{B}} g(x).$$

One reasons in an analogous way if g has a limit, and in the case of products.

7.4. Semicontinuous Functions

7.4.1. Definition. Let X be a topological space, $x_0 \in X$, and $f \in \mathscr{F}(X, \overline{\mathbf{R}})$. We say that f is *lower semicontinuous at* x_0 if, for every $\lambda < f(x_0)$, there exists a neighborhood V of x_0 in X such that

$$x \in V \Rightarrow f(x) \geq \lambda.$$

We say that f is *upper semicontinuous at* x_0 if, for every $\mu > f(x_0)$, there exists a neighborhood V of x_0 in X such that

$$x \in V \Rightarrow f(x) \leq \mu.$$

7.4.2. To say that f is lower semicontinuous at x_0 amounts to saying that $-f$ is upper semicontinuous at x_0. It therefore suffices, in principle, to study the properties of lower semicontinuous functions.

7.4.3. Example. Let X be a topological space, $x_0 \in X$, and $f \in \mathscr{F}(X, \overline{\mathbf{R}})$. Then: f is continuous at $x_0 \Leftrightarrow f$ is both lower and upper semicontinuous at x_0.

7.4.4. Example. For $x \in \mathbf{R}$, set $f(x) = x$ if $x \neq 0$, and $f(0) = 1$. Then f is continuous at every point of $\mathbf{R} - \{0\}$, and f is upper semicontinuous at 0 but not lower semicontinuous there.

7.4.5. Definition. Let X be a set, $(f_i)_{i \in I}$ a family of numerical functions on X. We denote by $\sup_{i \in I} f_i$, $\inf_{i \in I} f_i$ the functions $x \mapsto \sup_{i \in I} f_i(x)$, $x \mapsto \inf_{i \in I} f_i(x)$

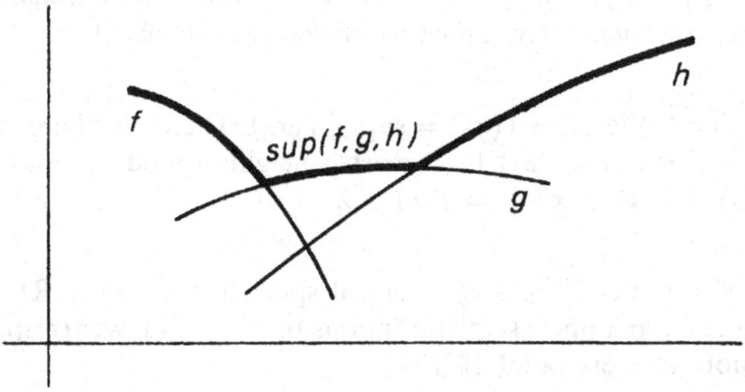

on X. These are called the *upper envelope* and *lower envelope* of the family $(f_i)_{i \in I}$. In particular, if f and g are two numerical functions on X, one can speak of $\sup(f, g)$ and $\inf(f, g)$.

7.4.6. Theorem. *Let* X *be a topological space,* $x_0 \in X$, f *and* g *numerical functions on* X *that are lower semicontinuous at* x_0. *Then* $\sup(f, g)$, $\inf(f, g)$ *and* $f + g$ *are lower semicontinuous at* x_0. *The same is true of* fg *if* $f \geq 0$ *and* $g \geq 0$.

We give the proof for $f + g$ (the other cases may be treated in an analogous way). If $f(x_0) = -\infty$ or $g(x_0) = -\infty$, the result is evident. Otherwise, let $\lambda < f(x_0) + g(x_0)$. There exist $\mu, \nu \in \mathbf{R}$ such that $\mu + \nu = \lambda$, $\mu < f(x_0)$, $\nu < g(x_0)$. Next, there exist neighborhoods V, W of x_0 in X such that

$$x \in V \Rightarrow f(x) \geq \mu, \qquad x \in W \Rightarrow g(x) \geq \nu.$$

Then

$$x \in V \cap W \Rightarrow f(x) + g(x) \geq \mu + \nu = \lambda.$$

7.4.7. The assertion 7.4.6 may be extended, one step at a time, to *finite* families of numerical functions.

7.4.8. The case of infinite families requires some precautions, as the following example shows. For $n = 1, 2, \ldots$, let f_n be the numerical function on $\mathbf{R}$ defined by $f_n(x) = e^{-nx^2}$. Each f_n is continuous. However, $\inf(f_n)$ is the function f on $\mathbf{R}$ such that $f(x) = 0$ for $x \neq 0$, $f(0) = 1$; and f is not lower semicontinuous at 0.

7.4.9. Nevertheless, we have the following result:

► **Theorem.** *Let* X *be a topological space,* $x_0 \in X$, $(f_i)_{i \in I}$ *a family of numerical functions on* X, *and* $f = \sup_{i \in I} f_i$. *If the* f_i *are lower semicontinuous at* x_0 *(for example, continuous at* x_0*), then* f *is lower semicontinuous at* x_0.

Let $\lambda < f(x_0)$. We have $f(x_0) = \sup_{i \in I} f_i(x_0)$. Therefore there exists an $i \in I$ such that $\lambda < f_i(x_0)$. Next, there exists a neighborhood V of x_0 such that $x \in V \Rightarrow f_i(x) \geq \lambda$. Then $x \in V \Rightarrow f(x) \geq \lambda$.

7.4.10. Definition. Let X be a topological space and $f \in \mathscr{F}(X, \overline{\mathbf{R}})$. We say that f is lower (resp. upper) semicontinuous on X if f is lower (resp. upper) semicontinuous at every point of X.

► **7.4.11. Theorem.** *Let* X *be a topological space, and* $f \in \mathscr{F}(X, \overline{\mathbf{R}})$. *The following conditions are equivalent*:

(i) *f is lower semicontinuous on* X;
(ii) *for every* $\lambda \in \overline{\mathbf{R}}$, *the set of* $x \in X$ *such that* $f(x) \leq \lambda$ *is closed*;
(iii) *for every* $\lambda \in \overline{\mathbf{R}}$, *the set of* $x \in X$ *such that* $f(x) > \lambda$ *is open.*

Since $f^{-1}([-\infty, \lambda])$ and $f^{-1}((\lambda, +\infty])$ are complementary in X, conditions (ii) and (iii) are equivalent.

(i) ⇒ (iii). Suppose f is lower semicontinuous on X. Let

$$A = f^{-1}((\lambda, +\infty]).$$

If $x_0 \in A$ then $f(x_0) > \lambda$; since f is lower semicontinuous at x_0, there exists a neighborhood V of x_0 such that $x \in V \Rightarrow f(x) > \lambda$. Therefore $V \subset A$. Thus, A is a neighborhood of each of its points, consequently is open.

(iii) ⇒ (i). Suppose condition (iii) is satisfied. Let $x_0 \in X$. Let $\lambda < f(x_0)$. The set $A = f^{-1}((\lambda, +\infty])$ is open, and $x_0 \in A$, thus A is a neighborhood of x_0. Since $f(x) > \lambda$ for all $x \in A$, we see that f is lower semicontinuous at x_0.

7.4.12. There are analogous characterizations of upper semicontinuous functions.

7.4.13. Corollary. *Let* X *be a topological space,* Y *a subset of* X, φ *the characteristic function of* Y *in* X. *For* Y *to be open* (resp. *closed*), *it is necessary and sufficient that* φ *be lower* (resp. *upper*) *semicontinuous.*

Let X_λ be the set of $x \in X$ such that $\varphi(x) > \lambda$. If $\lambda < 0$, then $X_\lambda = X$. If $0 \leq \lambda < 1$, then $X_\lambda = Y$. If $\lambda \geq 1$, then $X_\lambda = \varnothing$. The sets X and $\varnothing$ are open in X. By 7.4.11, we thus have

$$\varphi \text{ lower semicontinuous} \Leftrightarrow Y \text{ open}.$$

From this, one deduces the characterization of the closed sets.

7.4.14. Theorem. *Let* X *be a compact space, f a lower semicontinuous function on* X, *and* $m = \inf_{x \in X} f(x)$. *There exists an* $x_0 \in X$ *such that* $f(x_0) = m$.

For every $\lambda > m$, let X_λ be the set of $x \in X$ such that $f(x) \leq \lambda$. This set is closed (7.4.11), and it is nonempty by the definition of the infimum m. Every finite intersection

$$X_{\lambda_1} \cap \cdots \cap X_{\lambda_n}$$

is nonempty (because if λ_i is the smallest of the numbers $\lambda_1, \ldots, \lambda_n$, then $X_{\lambda_1} \cap \cdots \cap X_{\lambda_n} = X_{\lambda_i}$). Since X is compact, $\bigcap_{\lambda > m} X_\lambda$ is nonempty. Let x_0 be a point of this intersection. For every $\lambda > m$, we have $x_0 \in X_\lambda$, that is, $f(x_0) \leq \lambda$. Therefore $f(x_0) \leq m$. Also $f(x_0) \geq m$, thus $f(x_0) = m$.

▶ **7.4.15. Theorem.** *Let* X *be a complete metric space,* $(f_i)_{i\in I}$ *a family of lower semicontinuous functions on* X *such that* $\sup_{i\in I} f_i(x) < +\infty$ *for each* $x \in X$. *Then there exist a nonempty open subset* U *of* X *and a finite number* M *such that* $f_i(x) \le M$ *for all* $i \in I$ *and* $x \in U$.

Let $f = \sup_{i\in I} f_i$. Then $f(x) < +\infty$ for all $x \in X$ by hypothesis, and f is lower semicontinuous by 7.4.9. For $n = 1, 2, \ldots$, let U_n be the set of $x \in X$ such that $f(x) > n$. This is an open subset of X (7.4.11). If all of the U_n were dense in X, there would exist a point x_0 belonging to all of the U_n (5.5.12). One would then have $f(x_0) > n$ for all n, therefore $f(x_0) = +\infty$, which is absurd. Thus for some integer n_0, U_{n_0} is not dense; in other words, there exists a nonempty open subset U of X disjoint from U_{n_0}. For all $x \in U$, we have $x \notin U_{n_0}$, therefore $f(x) \le n_0$, consequently $f_i(x) \le n_0$ for all $i \in I$.

7.5. Stone–Weierstrass Theorem

7.5.1. Lemma. *Let* X *be a compact space,* $\mathscr{H}$ *a subset of* $\mathscr{C}(X, \mathbf{R})$ *having the following properties:*

(i) *if* $u \in \mathscr{H}$ *and* $v \in \mathscr{H}$, *then* $\sup(u, v) \in \mathscr{H}$ *and* $\inf(u, v) \in \mathscr{H}$;
(ii) *if* x, y *are points of* X *and if* $\alpha, \beta \in \mathbf{R}$ (*with* $\alpha = \beta$ *if* $x = y$), *then there exists* $u \in \mathscr{H}$ *such that* $u(x) = \alpha$, $u(y) = \beta$.

Then every function in $\mathscr{C}(X, \mathbf{R})$ *is the uniform limit of a sequence of functions in* $\mathscr{H}$.

Let $f \in \mathscr{C}(X, \mathbf{R})$ and $\varepsilon > 0$. We are to construct $g \in \mathscr{H}$ such that $f - \varepsilon \le g \le f + \varepsilon$.

(a) Let $x_0 \in X$. Let us show that there exists a function $u \in \mathscr{H}$ such that $u(x_0) = f(x_0)$ and $u \ge f - \varepsilon$.

For every $y \in X$, there exists a function $u_y \in \mathscr{H}$ such that $u_y(x_0) = f(x_0)$ and $u_y(y) = f(y)$. The set V_y of all $x \in X$ such that $u_y(x) > f(x) - \varepsilon$ is open (cf. 2.4.5). We have $y \in V_y$, thus the V_y, as y runs over X, form an open covering of X. Since X is compact, it is covered by sets $V_{y_1}, \ldots, V_{y_n}$. Let

$$u = \sup(u_{y_1}, u_{y_2}, \ldots, u_{y_n}) \in \mathscr{H}.$$

We have $u_{y_i}(x_0) = f(x_0)$ for all i, therefore $u(x_0) = f(x_0)$. Let $x \in X$. We have $x \in V_{y_i}$ for some i. Then $u(x) \ge u_{y_i}(x) > f(x) - \varepsilon$, and u satisfies the stated conditions.

(b) The function u constructed in (a) depends on x_0. For every $x \in X$, let us define similarly a function $v_x \in \mathscr{H}$ such that $v_x(x) = f(x)$ and $v_x \ge f - \varepsilon$. The set W_x of all $z \in X$ such that $v_x(z) < f(z) + \varepsilon$ is open. We have $x \in W_x$. Since X is compact, it is covered by sets $W_{x_1}, \ldots, W_{x_p}$. Let

$$g = \inf(v_{x_1}, \ldots, v_{x_p}) \in \mathscr{H}.$$

We have $v_{x_i} \geq f - \varepsilon$ for all i, therefore $g \geq f - \varepsilon$. Let $x \in X$. We have $x \in W_{x_j}$ for some j. Then $g(x) \leq v_{x_j}(x) < f(x) + \varepsilon$. Thus $g \leq f + \varepsilon$.

7.5.2. Lemma. *The function $\sqrt{t}$ on $[0, 1]$ is the uniform limit of a sequence of polynomials in t with real coefficients.*

We define functions $p_0(t), p_1(t), p_2(t), \ldots$, for $t \in [0, 1]$, recursively, in the following way:

$$p_0(t) = 0,$$
$$p_{n+1}(t) = p_n(t) + \tfrac{1}{2}(t - p_n(t)^2).$$

Let us show by induction that

$$p_0(t), p_1(t), \ldots, p_n(t) \text{ are polynomials in } t,$$

and

$$0 \leq p_0(t) \leq p_1(t) \leq \cdots \leq p_n(t) \leq \sqrt{t} \text{ on } [0, 1].$$

This is certainly the case for $n = 0$. Let us admit the preceding statements and let us prove the corresponding results for $n + 1$. It is first of all immediate that $p_{n+1}(t)$ is a polynomial in t. Next, for $t \in [0, 1]$, we have $t \geq p_n(t)^2$, therefore $p_{n+1}(t) \geq p_n(t)$. Finally,

$$\begin{aligned} p_{n+1}(t) - \sqrt{t} &= p_n(t) - \sqrt{t} + \tfrac{1}{2}(t - p_n(t)^2) \\ &= (p_n(t) - \sqrt{t})(1 - \tfrac{1}{2}(p_n(t) + \sqrt{t})). \end{aligned}$$

Now, $p_n(t) + \sqrt{t} \leq 2\sqrt{t}$, therefore $1 - \tfrac{1}{2}(p_n(t) + \sqrt{t}) \geq 1 - \sqrt{t} \geq 0$ in $[0, 1]$, and $p_n(t) - \sqrt{t} \leq 0$, therefore $p_{n+1}(t) \leq \sqrt{t}$.

For every $t \in [0, 1]$, the sequence $(p_n(t))$ is increasing and bounded above by $\sqrt{t}$, therefore has a finite limit $f(t) \geq 0$ that satisfies

$$f(t) = f(t) + \tfrac{1}{2}(t - f(t)^2),$$

whence $f(t) = \sqrt{t}$. Finally, the p_n tend to f uniformly on $[0, 1]$ by 6.2.9.

► **7.5.3. Theorem** (Stone–Weierstrass). *Let X be a compact space, $\mathscr{H}$ a subset of $\mathscr{C}(X, \mathbf{R})$ having the following properties:*

(i) *the constant functions belong to $\mathscr{H}$;*
(ii) *if $u, v \in \mathscr{H}$, then $u + v \in \mathscr{H}$ and $uv \in \mathscr{H}$;*
(iii) *if x, y are distinct points of X, there exists $u \in \mathscr{H}$ such that $u(x) \neq u(y)$.*

Then every function in $\mathscr{C}(X, \mathbf{R})$ is the uniform limit of a sequence of functions in $\mathscr{H}$.

Let $\overline{\mathscr{H}}$ be the closure of $\mathscr{H}$ in $\mathscr{C}(X, \mathbf{R})$ for the topology of uniform convergence. We are going to show that $\overline{\mathscr{H}}$ possesses the properties (i) and (ii) of 7.5.1. Then, by 7.5.1, the closure of $\overline{\mathscr{H}}$ will be equal to $\mathscr{C}(X, \mathbf{R})$, whence $\overline{\mathscr{H}} = \mathscr{C}(X, \mathbf{R})$, which will imply the theorem.

(a) If $u \in \overline{\mathscr{H}}$ and $v \in \overline{\mathscr{H}}$, then $u + v \in \overline{\mathscr{H}}$. For, there exist sequences (u_n), (v_n) of functions in $\mathscr{H}$ such that $u_n \to u$, $v_n \to v$ uniformly; then $u_n + v_n \to u + v$ uniformly, whence $u + v \in \overline{\mathscr{H}}$. Similarly, $uv \in \overline{\mathscr{H}}$; and if $\lambda \in \mathbf{R}$, then $\lambda u \in \overline{\mathscr{H}}$. Thus, every polynomial in u, that is, every function of the form

$$\lambda_0 + \lambda_1 u + \lambda_2 u^2 + \cdots + \lambda_n u^n,$$

where $\lambda_0, \ldots, \lambda_n \in \mathbf{R}$, belongs to $\overline{\mathscr{H}}$.

(b) Let x, y be distinct points of X, and α, $\beta \in \mathbf{R}$. There exists $v \in \mathscr{H}$ such that $v(x) \neq v(y)$. Set

$$v' = \frac{1}{v(x) - v(y)}(v - v(y)).$$

Then $v' \in \mathscr{H}$, $v'(x) = 1$, $v'(y) = 0$. Let

$$u = \beta + (\alpha - \beta)v'.$$

Then $u \in \mathscr{H}$ and $u(x) = \alpha$, $u(y) = \beta$.

(c) Let $u \in \overline{\mathscr{H}}$ and let us show that $|u| \in \overline{\mathscr{H}}$. The function u, being continuous on X, is bounded (4.2.13). On multiplying u by a suitable constant, we are thus reduced to the case that $-1 \leq u \leq 1$. Then $0 \leq u^2 \leq 1$. Let $\varepsilon > 0$. By 7.5.2, there exists a polynomial $p(t)$ with real coefficients, such that $|p(t) - \sqrt{t}| \leq \varepsilon$ for all $t \in [0, 1]$. Then $|p(u(x)^2) - \sqrt{u(x)^2}| \leq \varepsilon$ for all $x \in X$, that is, $\big|p(u^2) - |u|\big| \leq \varepsilon$. Now, $p(u^2) \in \overline{\mathscr{H}}$ by (a). Thus $|u|$ is adherent to $\overline{\mathscr{H}}$, consequently $|u| \in \overline{\mathscr{H}}$.

(d) Let $u, v \in \overline{\mathscr{H}}$. In view of (a) and (c), we have

$$\sup(u, v) = \tfrac{1}{2}(u + v + |u - v|) \in \overline{\mathscr{H}},$$

$$\inf(u, v) = \tfrac{1}{2}(u + v - |u - v|) \in \overline{\mathscr{H}}.$$

7.5.4. Corollary. *Let* X *be a compact space,* $\mathscr{H}$ *a set of continuous, complex-valued functions on* X, *having the following properties*:

(i) *the complex-valued constant functions belongs to* $\mathscr{H}$;
(ii) *if* $u, v \in \mathscr{H}$, *then* $u + v \in \mathscr{H}$, $uv \in \mathscr{H}$ *and* $\bar{u} \in \mathscr{H}$;
(iii) *if* x, y, *are distinct points of* X, *there exists* $u \in \mathscr{H}$ *such that* $u(x) \neq u(y)$.

Then every function in $\mathscr{C}(X, \mathbf{C})$ *is the uniform limit of a sequence of functions in* $\mathscr{H}$.

Let $\mathscr{H}'$ be the set of functions belonging to $\mathscr{H}$ that are real-valued. Then $\mathscr{H}'$ satisfies the conditions (i) and (ii) of 7.5.3. If x, y are distinct points of X, there exists $u \in \mathscr{H}$ such that $u(x) \neq u(y)$. Then either $\operatorname{Re} u(x) \neq \operatorname{Re} u(y)$ or $\operatorname{Im} u(x) \neq \operatorname{Im} u(y)$. Now,

$$\operatorname{Re} u = \tfrac{1}{2}(u + \bar{u}) \in \mathscr{H}' \quad \text{and} \quad \operatorname{Im} u = \frac{1}{2i}(u - \bar{u}) \in \mathscr{H}',$$

therefore $\mathscr{H}'$ also satisfies condition (iii) of 7.5.3. Let $g \in \mathscr{C}(X, \mathbf{C})$. Then $g = g_1 + ig_2$ with $g_1, g_2 \in \mathscr{C}(X, \mathbf{R})$. By 7.5.3, g_1 and g_2 are uniform limits of functions in $\mathscr{H}'$, therefore g is the uniform limit of functions in $\mathscr{H}$.

7.5.5. Corollary. *Let X be a compact subset of* $\mathbf{R}^n$, *and* $f \in \mathscr{C}(X, \mathbf{C})$. *Then f is the uniform limit on X of a sequence of polynomials in n variables with complex coefficients.*

Consider the polynomials in n variables with complex coefficients. These are functions on $\mathbf{R}^n$ whose restrictions to X form a subset $\mathscr{H}$ of $\mathscr{C}(X, \mathbf{C})$. It is clear that $\mathscr{H}$ satisfies conditions (i), (ii), (iii) of 7.5.4, whence the corollary.

7.5.6. Corollary. *Let f be a continuous, complex-valued periodic function on* $\mathbf{R}$ *of period* 1. *Then f is the uniform limit on* $\mathbf{R}$ *of a sequence of trigonometric polynomials (that is, functions of the form*

$$t \mapsto \sum_{r=-n}^{n} a_r e^{2\pi irt},$$

where the a_r *are complex constants).*

Let p be the canonical mapping of $\mathbf{R}$ onto $\mathbf{T}$ (3.4.3). Since f has period 1, there exists a complex-valued function g on $\mathbf{T}$ such that $f(x) = g(p(x))$ for all $x \in \mathbf{R}$. By 3.4.4, g is continuous. By 4.2.16, there exists a homeomorphism θ of $\mathbf{U}$ onto $\mathbf{T}$ such that $\theta^{-1}(p(x)) = e^{2\pi ix}$ for all $x \in \mathbf{R}$. Let $h = g \circ \theta \in \mathscr{C}(\mathbf{U}, \mathbf{C})$. For all $x \in \mathbf{R}$, $f(x) = g(p(x)) = h(\theta^{-1}(p(x))) = h(e^{2\pi ix})$.

Now, $\mathbf{U}$ is a compact subset of $\mathbf{R}^2 = \mathbf{C}$. Let $\varepsilon > 0$. There exists a polynomial $\sum_{m,n} a_{mn} x^m y^n$ in x and y, with complex coefficients, such that

$$\left| h(x + iy) - \sum_{m,n} a_{mn} x^m y^n \right| \leq \varepsilon$$

for every point $x + iy$ of $\mathbf{U}$ (7.5.5). Consequently, for every $t \in \mathbf{R}$, we have

$$\left| h(e^{2\pi it}) - \sum_{m,n} a_{mn} (\cos 2\pi t)^m (\sin 2\pi t)^n \right| \leq \varepsilon,$$

that is,

$$\left| f(t) - \sum_{m,n} a_{mn} (\cos 2\pi t)^m (\sin 2\pi t)^n \right| \leq \varepsilon.$$

Since $\cos 2\pi t = \frac{1}{2}(e^{2\pi it} + e^{-2\pi it})$ and $\sin 2\pi t = (1/2i)(e^{2\pi it} - e^{-2\pi it})$, the function

$$\sum_{m,n} a_{mn} (\cos 2\pi t)^m (\sin 2\pi t)^n$$

is a trigonometric polynomial.

* **7.5.7.** Let X be a noncompact locally compact space. The sets of the form $X - C$, where C is compact in X, form a filter base $\mathscr{B}$ on X (4.2.9(i)). Let $X' = X \cup \{\omega\}$ be the Alexandroff compactification of X (4.5.9). By 4.5.8, the sets $(X - C) \cup \{\omega\}$, where C is compact in X, are the open neighborhoods of ω in X'.

Consequently, if f is a complex-valued function on X, the following conditions are equivalent:

(i) f tends to 0 along $\mathscr{B}$;
(ii) if f' is the function on X' that extends f and vanishes at ω, then $\lim_{x\to\omega} f'(x) = 0$.

When these conditions are satisfied, we say that f *tends to* 0 *at infinity* on X. We remark that if, in addition, f is continuous on X, then f' is continuous on X'.

* **7.5.8. Corollary.** *Let* X *be a noncompact locally compact space,* $\mathscr{C}_0$ *the set of continuous complex-valued functions on* X *that tend to* 0 *at infinity, and* $\mathscr{H}$ *a subset of* $\mathscr{C}_0$ *having the following properties:*

(i) *if* $u, v \in \mathscr{H}$ *and* $\lambda \in \mathbf{C}$, *then* $u + v \in \mathscr{H}$, $uv \in \mathscr{H}$, $\bar{u} \in \mathscr{H}$ *and* $\lambda u \in \mathscr{H}$;
(ii) *if* x, y *are distinct points of* X, *there exists* $u \in \mathscr{H}$ *such that* $u(x) \neq u(y)$;
(iii) *if* $x \in X$, *there exists* $u \in \mathscr{H}$ *such that* $u(x) \neq 0$.

Then every function in $\mathscr{C}_0$ *is the uniform limit of a sequence of functions in* $\mathscr{H}$.

Let us keep the notations of 7.5.7. Let $\mathscr{H}'$ be the set of functions on X' of the form $f' + \lambda$, where $f \in \mathscr{H}$ and $\lambda \in \mathbf{C}$. Then $\mathscr{H}' \subset \mathscr{C}(X', \mathbf{C})$. Obviously $\mathscr{H}'$ satisfies condition (i) of 7.5.4. Let $u, v \in \mathscr{H}'$. We have $u = f' + \lambda$, $v = g' + \mu$ with $f, g \in \mathscr{H}$ and $\lambda, \mu \in \mathbf{C}$. Then:

$$\begin{aligned} u + v &= f' + g' + \lambda + \mu = (f + g)' + (\lambda + \mu) \in \mathscr{H}', \\ uv &= f'g' + \lambda g' + \mu f' + \lambda\mu \\ &= (fg + \lambda g + \mu f)' + \lambda\mu \in \mathscr{H}', \end{aligned}$$

and

$$\bar{u} = \bar{f}' + \bar{\lambda} = (\bar{f})' + \bar{\lambda} \in \mathscr{H}'.$$

Finally, if x, y are distinct points of X', there exists $u \in \mathscr{H}'$ such that $u(x) \neq u(y)$ (if $x, y \in X$, this follows from the hypothesis (ii); if $x = \omega$ or $y = \omega$, it follows from the hypothesis (iii)). Now let $h \in \mathscr{C}_0$ and $\varepsilon > 0$. By 7.5.4, there exist $f \in \mathscr{H}$ and $\lambda \in \mathbf{C}$ such that

$$|h'(x) - f'(x) - \lambda| \leq \frac{\varepsilon}{2}$$

for all $x \in X'$. In particular, $\varepsilon/2 \geq |h'(\omega) - f'(\omega) - \lambda| = |\lambda|$. Therefore

$$|h(x) - f(x)| \leq \frac{\varepsilon}{2} + \frac{\varepsilon}{2} = \varepsilon$$

for all $x \in X$.

* 7.6. Normal Spaces

7.6.1. Theorem. *Let X be a topological space. The following conditions are equivalent:*

(i) *For any disjoint closed subsets A and B of X, there exist disjoint open sets U and V of X such that* $A \subset U$, $B \subset V$.
(ii) *For every closed subset A of X and every open set W of X such that* $A \subset W$, *there exists an open set W′ of X such that* $A \subset W' \subset \overline{W'} \subset W$.
(iii) *For any disjoint closed subsets A and B of X, there exists a continuous mapping of X into* $[0, 1]$ *equal to 0 at every point of A and to 1 at every point of B.*
(iv) *For every closed subset A of X and every numerical function f defined and continuous on A, there exists a numerical function defined and continuous on X that extends f.*

(iv) ⇒ (i). Suppose that condition (iv) is satisfied. Let A and B be disjoint closed subsets of X. Then $C = A \cup B$ is a closed subset of X. Set $f(x) = 0$ for $x \in A$ and $f(x) = 1$ for $x \in B$. Then f is continuous on C (because A and B are open in C). By (iv), there exists a continuous mapping g of X into $\overline{\mathbf{R}}$ that extends f. Let

$$U = g^{-1}((-\infty, \tfrac{1}{2})), \qquad V = g^{-1}((\tfrac{1}{2}, +\infty)).$$

Then U and V are disjoint open sets in X, and $A \subset U$, $B \subset V$.

(i) ⇒ (ii). Suppose that condition (i) is satisfied. Let A (resp. W) be a closed (resp. open) set in X with $A \subset W$. Set $B = X - W$; this is a closed set in X disjoint from A. By (i), there exist disjoint open sets U and V of X such that $A \subset U$, $B \subset V$. Then $U \subset X - V$ and $X - V$ is closed, therefore $\overline{U} \subset X - V \subset X - B = W$. Thus $A \subset U \subset \overline{U} \subset W$.

(ii) ⇒ (iii). Suppose that condition (ii) is satisfied. Let A and B be disjoint closed subsets of X. We are to construct a continuous mapping of X into $[0, 1]$ equal to 0 on A and to 1 on B.

Let D be the set of 'dyadic' numbers belonging to $[0, 1]$, that is, the set of numbers of the form $k/2^n$, where $n = 0, 1, 2, \ldots$ and $k = 0, 1, 2, \ldots, 2^n$. This set is dense in $[0, 1]$. For every $a \in D$, we are going to construct an open subset $U(d)$ of X in such a way that

(1) $$d < d' \Rightarrow \overline{U(d)} \subset U(d').$$

We set $U(1) = X - B$ and we choose an open set $U(0)$ such that

$$A \subset U(0) \subset \overline{U(0)} \subset U(1) \tag{2}$$

(this is possible by (ii)). Suppose the $U(d)$ already chosen for $d = 0, 1/2^n, 2/2^n, \ldots, 2^n/2^n = 1$, in such a way that

$$\overline{U(k/2^n)} \subset U((k+1)/2^n) \quad \text{for} \quad 0 \leq k < 2^n.$$

Let us define $U(k/2^{n+1})$ for $k = 0, 1, \ldots, 2^{n+1}$. For k even, $U(k/2^{n+1})$ has already been chosen. For k odd, thus of the form $2h + 1$, we choose an open set $U((2h+1)/2^{n+1})$ such that

$$\overline{U(2h/2^{n+1})} \subset U((2h+1)/2^{n+1}) \subset \overline{U((2h+1)/2^{n+1})} \subset U((2h+2)/2^{n+1}),$$

which is possible by (ii). By induction, the open sets $U(d)$ are thus defined for all $d \in D$, and property (1) certainly holds.

If $x \in B$ we set $f(x) = 1$. If $x \notin B$, then $x \in U(1)$; let $f(x)$ be the infimum in $\mathbf{R}$ of the $d \in D$ such that $x \in U(d)$. We have thus defined a mapping f of X into $[0, 1]$, equal to 1 at every point of B. If $x \in A$ then $x \in U(0)$ by (2), therefore $f(x) = 0$.

Finally, let us show that f is continuous at every point x of X. Let $a = f(x)$ and $\varepsilon > 0$.

If $0 < a < 1$, then there exist $d, d', d'' \in D$ such that

$$a - \varepsilon \leq d < d' < a < d'' \leq a + \varepsilon.$$

If one had $x \notin U(d'')$, it would follow that $f(x) \geq d''$, which is absurd; thus $x \in U(d'')$. On the other hand, $f(x) > d'$, therefore $x \notin U(d')$, therefore $x \notin \overline{U(d)}$ by (1). Consequently, if one sets $V = U(d'') \cap (X - \overline{U(d)})$, V is an open neighborhood of x. Let $y \in V$. Then $y \in U(d'')$, therefore $f(y) \leq d''$; and $f(y) \geq d$, since otherwise one would have $y \in U(d)$. Thus,

$$y \in V \Rightarrow |f(y) - f(x)| \leq \varepsilon,$$

which proves our assertion when $0 < a < 1$.

If $a = 1$, there exist $d, d' \in D$ such that $a - \varepsilon \leq d < d' < a = 1$. We see as above that $V = X - \overline{U(d)}$ is an open neighborhood of x and that, for every $y \in V$, one has $f(y) \geq d$, therefore $|f(y) - f(x)| \leq \varepsilon$.

If $a = 0$, there exists $d'' \in D$ such that $0 = a < d'' \leq a + \varepsilon$. One sees as above that $V = U(d'')$ is an open neighborhood of x and that, for every $y \in V$, one has $f(y) \leq d''$, therefore $|f(y) - f(x)| \leq \varepsilon$.

(iii) $\Rightarrow$ (iv). Suppose that condition (iii) is satisfied. Let A be a closed set in X and let f be a continuous mapping of A into $\overline{\mathbf{R}}$. Let us define a continuous mapping of X into $\overline{\mathbf{R}}$ that extends f. Since $\overline{\mathbf{R}}$ and $[-1, 1]$ are homeomorphic, we can suppose that f takes its values in $[-1, 1]$; we will define a continuous mapping of X into $[-1, 1]$ that extends f.

We shall first prove the following intermediary result:

(*) If u is a continuous mapping of A into $[-1, 1]$, there exists a continuous mapping v of X into $[-\frac{1}{3}, \frac{1}{3}]$ such that $|u(x) - v(x)| \leq \frac{2}{3}$ for all $x \in A$.

For, let H (resp. K) be the set of $x \in A$ such that $-1 \leq u(x) \leq -\frac{1}{3}$ (resp. $\frac{1}{3} \leq u(x) \leq 1$). The sets H and K are closed in A, therefore in X, and they are disjoint. By (iii), there exists a continuous mapping v of X into $[-\frac{1}{3}, \frac{1}{3}]$, equal to $-\frac{1}{3}$ at every point of H and to $\frac{1}{3}$ at every point of K. It is clear that $|u(x) - v(x)| \leq \frac{2}{3}$ for all $x \in A$.

This established, we are going to recursively construct continuous numerical functions $g_0, g_1, g_2, \ldots$ on X such that

$$(3) \qquad \begin{cases} -1 + (\frac{2}{3})^{n+1} \leq g_n(x) \leq 1 - (\frac{2}{3})^{n+1} & \text{for all} \quad x \in X \\ |f(x) - g_n(x)| \leq (\frac{2}{3})^{n+1} & \text{for all} \quad x \in A. \end{cases}$$

The existence of g_0 results from applying (*) with $u = f$. Suppose g_0, $g_1, \ldots, g_n$ already constructed. Define

$$u(x) = (\tfrac{3}{2})^{n+1}(f(x) - g_n(x)) \quad \text{for} \quad x \in A.$$

Then u is a continuous mapping of A into $[-1, 1]$. By (*), there exists a continuous mapping v of X into $[-\frac{1}{3}, \frac{1}{3}]$ such that $|u(x) - v(x)| \leq \frac{2}{3}$ for all $x \in A$. Then, for all $x \in A$,

$$|f(x) - g_n(x) - (\tfrac{2}{3})^{n+1}v(x)| = (\tfrac{2}{3})^{n+1}|u(x) - v(x)| \leq (\tfrac{2}{3})^{n+2}.$$

On the other hand, for all $x \in X$,

$$|g_n(x) + (\tfrac{2}{3})^{n+1}v(x)| \leq 1 - (\tfrac{2}{3})^{n+1} + (\tfrac{2}{3})^{n+1} \cdot \tfrac{1}{3} = 1 - (\tfrac{2}{3})^{n+2}.$$

Setting

$$g_{n+1}(x) = g_n(x) + (\tfrac{2}{3})^{n+1}v(x)$$

for every $x \in X$, the construction of the g_n is complete by induction. We observe, moreover, that

$$|g_{n+1}(x) - g_n(x)| \leq \tfrac{1}{3}(\tfrac{2}{3})^{n+1} \quad \text{for all} \quad x \in X.$$

Since the series with general term $\frac{1}{3}(\frac{2}{3})^{n+1}$ is convergent, the series with general term $g_{n+1} - g_n$ is normally convergent, therefore uniformly convergent (6.1.10); in other words, g_n has a uniform limit g on X. This limit is continuous (6.1.11). By (3), we have $-1 \leq g(x) \leq 1$ for all $x \in X$ and $f(x) = g(x)$ for all $x \in A$.

7.6.2. Definition. One calls *normal space* a separated space that satisfies the equivalent conditions of 7.6.1.

7.6.3. Examples. (a) *Every metric space is normal.* For let X be a metric space, d its metric, A and B disjoint, nonempty closed subsets of X. Since the functions $x \mapsto d(x, A)$ and $x \mapsto d(x, B)$ are continuous on X (5.1.6), the set U (resp. V) of $x \in X$ such that $d(x, A) < d(x, B)$ (resp. $d(x, B) < d(x, A)$) is open in X. It is clear that $U \cap V = \emptyset$. If $x \in A$ then $d(x, A) = 0$ and $d(x, B) > 0$ (because the relation $d(x, B) = 0$ would imply $x \in \bar{B}$, therefore $x \in B$). Thus $A \subset U$, and similarly $B \subset V$.

(b) *Every compact space is normal.* This follows from 4.2.11(i).

(c) If I is an uncountable set, it can be shown that $\mathbf{R}^I$ is not normal. One can also construct locally compact spaces that are not normal.

7.6.4. Remark. Let X be a normal space, A a closed subset of X, f a numerical function defined and continuous on A. We know that there exists a numerical function g, defined and continuous on X, that extends f. Suppose, moreover, that f is *finite*. We shall see that g can then be chosen to be finite.

Suppose first that $f \geq 0$ on A. *A fortiori*, f takes its values in $[0, +\infty]$, therefore we can suppose that g takes its values in $[0, +\infty]$ (which is homeomorphic to $\overline{\mathbf{R}}$). Let $B = g^{-1}(\{+\infty\})$. Then B is closed and is disjoint from A. The function h on the closed set $A \cup B$ that is equal to f on A and to 0 on B is therefore continuous. Let g' be a continuous extension of h to X taking its values in $[0, +\infty]$. Replacing g by $\inf(g, g')$, we obtain a continuous ex-extension of f to X that is finite at every point of X.

In the general case, let $f_1 = \sup(f, 0)$, $f_2 = \sup(-f, 0)$. Then $f = f_1 - f_2$, and f_1, f_2 are finite, ≥ 0 and continuous. It now suffices to apply the preceding paragraph to f_1 and f_2.

7.6.5. Theorem. *Let* X *be a compact space. There exists a set* I *such that* X *is homeomorphic to a closed subset of* $[0, 1]^I$.

Let $(f_i)_{i \in I}$ be a family of continuous mappings of X into $[0, 1]$. By 3.3.2(d), the mapping $f : x \mapsto (f_i(x))_{i \in I}$ of X into $[0, 1]^I$ is continuous. The set $f(X)$ is a compact, hence closed, subset of $[0, 1]^I$ (4.2.12, 4.2.7). If f is injective, then f is a homeomorphism of X onto $f(X)$ (4.2.15).

Now, if one takes for $(f_i)_{i \in I}$ the family of *all* continuous mappings of X into $[0, 1]$, then f is injective. For, let a and b be distinct points of X. Since X is normal (7.6.3(b)), there exists a continuous mapping g of X into $[0, 1]$ such that $g(a) = 0$, $g(b) = 1$. Since g is one of the f_i, we have $f(a) \neq f(b)$.

7.6.6. Theorem 7.6.5 is sometimes expressed by saying that every compact space may be *embedded* in a 'generalized cube'.

7.6.7. Theorem. *Let* X *and* Y *be compact spaces. Let* $\mathscr{H}$ *be the set of functions on* $X \times Y$ *of the form*

$$(x, y) \mapsto f_1(x)g_1(y) + f_2(x)g_2(y) + \cdots + f_n(x)g_n(y),$$

where $f_1, \ldots, f_n \in \mathscr{C}(X, \mathbf{R})$, $g_1, \ldots, g_n \in \mathscr{C}(Y, \mathbf{R})$, $n = 1, 2, \ldots$. *Then* $\mathscr{H}$ *is dense in* $\mathscr{C}(X \times Y, \mathbf{R})$ *for the topology of uniform convergence.*

To apply the Stone–Weierstrass theorem, it suffices to check that $\mathscr{H}$ satisfies the conditions (i), (ii), (iii) of 7.5.3 relative to the compact space $X \times Y$. This is clear for conditions (i) and (ii). Let (a, b) and (a', b') be two distinct points of $X \times Y$. Suppose for example that $a \neq a'$. Since X is normal, there exists $f \in \mathscr{C}(X, \mathbf{R})$ such that $f(a) = 0$, $f(a') = 1$. Set $g(y) = 1$ for all $y \in Y$. Then the function $(x, y) \mapsto f(x)g(y)$ on $X \times Y$ takes on different values at (a, b) and (a', b').

CHAPTER VIII

Normed Spaces

We take up again the theory of normed spaces and pre-Hilbert spaces. §§1 to 5 are already familiar, excepting possibly Theorem 8.3.4 on equivalent norms. In §§6, 7, 8 we make the connection between this theory and that of complete spaces; some of these results (Banach–Steinhaus theorem, Riesz's theorem) are very fruitful, but the reader can hardly be convinced of this unless (s)he studies 'functional analysis' later on.

8.1. Definition of Normed Spaces

8.1.1. Definition. Let E be a vector space over **R** or **C**. A *seminorm* on E is a function $x \mapsto \|x\|$ defined on E, with finite values ≥ 0, such that

(a) $\|\lambda x\| = |\lambda| \|x\|$ for all $x \in \mathrm{E}$ and all scalars λ;
(b) $\|x + y\| \leq \|x\| + \|y\|$ for all $x \in \mathrm{E}$ and $y \in \mathrm{E}$ (triangle inequality).

It follows from (a) that $x = 0 \Rightarrow \|x\| = 0$. If, conversely, $\|x\| = 0 \Rightarrow x = 0$, the seminorm is called a *norm*.

When a seminorm (resp. norm) is given on E, we say that E is a *seminormed* (resp. *normed*) vector space.

Conditions (a) and (b) imply at once:

$$\|-x\| = \|x\| \qquad \text{for all } x \in \mathrm{E},$$

$$\|x - y\| \leq \|x\| + \|y\| \quad \text{for all } x \in \mathrm{E} \text{ and } y \in \mathrm{E}.$$

There is an obvious notion of isomorphism between normed or seminormed spaces.

8.1.2. Example. In $\mathbf{R}^n$ or $\mathbf{C}^n$ one can define, for example, the following norms:

$$\|(x_1, \ldots, x_n)\| = (|x_1|^2 + \cdots + |x_n|^2)^{1/2},$$
$$\|(x_1, \ldots, x_n)\| = |x_1| + \cdots + |x_n|,$$
$$\|(x_1, \ldots, x_n)\| = \sup(|x_1|, \ldots, |x_n|).$$

8.1.3. Example. Let X be a set, E the vector space of all bounded real-valued (or complex-valued) functions defined on X. For every $f \in \mathrm{E}$, set

$$\|f\| = \sup_{x \in \mathrm{X}} |f(x)|.$$

One verifies immediately that $f \mapsto \|f\|$ is a norm on E.

Now let A be a subset of X. This time, for every $f \in \mathrm{E}$ set

$$\|f\| = \sup_{x \in \mathrm{A}} |f(x)|.$$

One verifies that $f \mapsto \|f\|$ is a seminorm on E.

8.1.4. Example. Let E be the vector space of all sequences $(\lambda_1, \lambda_2, \ldots)$ of real (or complex) numbers. The bounded sequences form a linear subspace of E, denoted $l^\infty_{\mathbf{R}}$ or $l^\infty_{\mathbf{C}}$ (or simply l^∞). For every $s = (x_1, x_2, \ldots) \in l^\infty$, set

$$\|s\| = \sup(|x_1|, |x_2|, \ldots).$$

Then l^∞ becomes a normed space. This is the special case of 8.1.3 where one takes $\mathrm{X} = \{1, 2, 3, \ldots\}$.

8.1.5. Example. We denote by $l^1_{\mathbf{C}}$ or $l^1_{\mathbf{R}}$ (or simply l^1) the set of sequences $s = (x_1, x_2, \ldots)$ of complex or real numbers such that $\sum_{n=1}^{\infty} |x_n| < +\infty$. This is a linear subspace of l^∞. For, if $s = (x_1, x_2, \ldots) \in l^1$ and $t = (y_1, y_2, \ldots) \in l^1$, then

$$(1) \qquad \sum_{n=1}^{\infty} |x_n + y_n| \leq \sum_{n=1}^{\infty} |x_n| + \sum_{n=1}^{\infty} |y_n| < +\infty,$$

thus $s + t \in l^1$. It is clear that $\lambda s \in l^1$ for every scalar λ.

For $s = (x_1, x_2, \ldots) \in l^1$, set

$$\|s\| = \sum_{n=1}^{\infty} |x_n|.$$

Then $s \mapsto \|s\|$ is a norm on l^1 (the triangle inequality results from (1)).

8.1.6. The Metric Deduced From a Norm. Let E be a normed vector space. For $x, y \in \mathrm{E}$, set $d(x, y) = \|x - y\|$. One verifies without difficulty that d is a metric on E. (Thus, a normed space is automatically a metric space, hence a topological space.) This metric is invariant under translations in E, that is, $d(x, y) = d(x + a, y + a)$ for all $x, y, a \in E$. For every $x \in E$, $d(x, 0) = \|x\|$.

8.1.7. Examples. Starting with the norms 8.1.2, one recovers the usual metrics on $\mathbf{R}^n$ or $\mathbf{C}^n$ (1.1.15). In l^∞, the distance between two sequences (x_i) and (y_i) is $\sup(|x_1 - y_1|, |x_2 - y_2|, \ldots)$. In l^1, the distance between two sequences (x_i) and (y_i) is $|x_1 - y_1| + |x_2 - y_2| + \cdots$. If f and g are two bounded, real-valued or complex-valued functions on a set X, then the distance between them deduced from the norm of 8.1.3 is

$$d(f, g) = \sup_{x \in X} |f(x) - g(x)|;$$

this is the metric of uniform convergence.

8.1.8. Theorem. *Let* E *be a normed space.*

(i) *The mapping* $(x, y) \mapsto x + y$ *of* $E \times E$ *into* E *is continuous.*
(ii) *The mapping* $(\lambda, x) \mapsto \lambda x$ *of* $\mathbf{R} \times E$ (*or* $\mathbf{C} \times E$) *into* E *is continuous.*

Let $x_0, y_0 \in E$ and $\varepsilon > 0$. If $x, y \in E$ are such that

$$\|x - x_0\| \leq \frac{\varepsilon}{2} \quad \text{and} \quad \|y - y_0\| \leq \frac{\varepsilon}{2},$$

then

$$\begin{aligned} \|(x + y) - (x_0 + y_0)\| &= \|(x - x_0) + (y - y_0)\| \\ &\leq \|x - x_0\| + \|y - y_0\| \leq \varepsilon. \end{aligned}$$

This proves (i).

Let $x_0 \in E$, $\lambda_0 \in \mathbf{R}$ (for example) and $\varepsilon > 0$. Set

$$\eta = \inf\left(1, \frac{\varepsilon}{1 + |\lambda_0| + \|x_0\|}\right) > 0.$$

Let $x \in E$, $\lambda \in \mathbf{R}$ be such that $|\lambda - \lambda_0| \leq \eta$ and $\|x - x_0\| \leq \eta$. Then

$$\begin{aligned} \|\lambda x - \lambda_0 x_0\| &= \|(\lambda - \lambda_0)(x - x_0) + \lambda_0(x - x_0) + (\lambda - \lambda_0)x_0\| \\ &\leq |\lambda - \lambda_0|\,\|x - x_0\| + |\lambda_0|\,\|x - x_0\| + |\lambda - \lambda_0|\,\|x_0\| \\ &\leq \eta^2 + |\lambda_0|\eta + \|x_0\|\eta \leq \eta(1 + |\lambda_0| + \|x_0\|) \leq \varepsilon. \end{aligned}$$

This proves (ii).

8.1.9. If E is a normed vector space, then every linear subspace of E, equipped with the restriction of the norm of E, is automatically a normed vector space. For example, let X be a topological space and F the set of continuous, bounded real-valued functions on X. Then F, equipped with the norm of uniform convergence, is a normed linear subspace of the normed space defined in 8.1.3.

8.1.10. Product of Seminormed Vector Spaces. Let $E_1, \ldots, E_n$ be seminormed vector spaces (all real or all complex). Let $E = E_1 \times \cdots \times E_n$, which is a real or complex vector space. For $x = (x_1, \ldots, x_n) \in E$, set

$$\|x\| = (\|x_1\|^2 + \cdots + \|x_n\|^2)^{1/2}.$$

One verifies in the usual way that this defines a seminorm on E. If $E_1, \ldots, E_n$ are normed spaces, then this seminorm is a norm, and the metric space defined by the norm of E is the product of the metric spaces $E_1, \ldots, E_n$ in the sense of 3.2.3.

Other useful norms can be defined on E; for example,

$$\|x\| = \|x_1\| + \cdots + \|x_n\|,$$

$$\|x\| = \sup(\|x_1\|, \ldots, \|x_n\|).$$

On $\mathbf{R}^n$ or $\mathbf{C}^n$, one recovers the norms of 8.1.2.

8.1.11. The Normed Space Associated with a Seminormed Space. Let E be a seminormed space. Let F be the set of $x \in E$ such that $\|x\| = 0$. If $x, y \in F$ then $\|x + y\| \le \|x\| + \|y\| = 0$, therefore $x + y \in F$. Obviously $\lambda x \in F$ for every scalar λ. Thus F is a linear subspace of E, and one can form the quotient vector space $E' = E/F$.

Let $x' \in E'$. Choose a representative x of x' in E. The number $\|x\|$ depends only on x' and not on the choice of the representative x. For, every other representative of x' is of the form $x + u$ with $u \in F$; then $\|x + u\| \le \|x\| + \|u\| = \|x\|$, and similarly $\|x\| \le \|x + u\| + \|u\| = \|x + u\|$, thus $\|x + u\| = \|x\|$. We may therefore set $\|x'\| = \|x\|$. Since $x \mapsto \|x\|$ is a seminorm on E, one verifies easily that $x' \mapsto \|x'\|$ is a seminorm on E'. *This seminorm is a norm*: for, if $x' \in E'$ is such that $\|x'\| = 0$, and if x is a representative of x' in E, then $\|x\| = 0$, therefore $x \in F$, therefore $x' = 0$. We say that *E' is the normed space associated with the seminormed space* E.

The study of the properties of E is practically equivalent to the study of the properties of E'.

8.2. Continuous Linear Mappings

8.2.1. Let E, F be normed spaces. Let u be a linear mapping of E into F. Let B be the closed ball in E with center 0 and radius 1. *We define*:

(1) $$\|u\| = \sup_{x \in B} \|ux\| \in [0, +\infty].$$

We have

(2) $$\|uy\| \le \|u\| \, \|y\| \quad \text{for all } y \in E$$

(with the convention that $0 \cdot +\infty = 0$). For, (2) is clear if $y = 0$. If $y \neq 0$, let $x = \|y\|^{-1}y$. Then $y = \|y\|x$, therefore $uy = \|y\|ux$, and so $\|uy\| = \|y\|\,\|ux\| \leq \|y\|\,\|u\|$ because $x \in \mathrm{B}$.

8.2.2. More precisely, $\|u\|$ is the smallest of the numbers $a \in [0, +\infty]$ such that

$$(3) \qquad \|uy\| \leq a\|y\| \quad \text{for all } y \in \mathrm{E}.$$

For, if a satisfies (3) then, in particular, $\|ux\| \leq a$ for all $x \in \mathrm{B}$, therefore $\|u\| \leq a$.

8.2.3. Let S be the sphere in E with center 0 and radius 1. If $\mathrm{E} \neq 0$, then every element of B may be written λx with $0 \leq \lambda \leq 1$ and $x \in \mathrm{S}$, therefore

$$(4) \qquad \|u\| = \sup_{x \in \mathrm{S}} \|ux\|.$$

► **8.2.4. Theorem.** *Let* E, F *be normed spaces and* u *a linear mapping of* E *into* F. *The following conditions are equivalent*:

(i) u *is continuous at* 0;
(ii) u *is continuous*;
(iii) u *is uniformly continuous*;
(iv) $\|u\| < +\infty$.

(iii) ⇒ (ii) ⇒ (i). This is obvious.

(i) ⇒ (iv). Suppose that u is continuous at 0. There exists an $\eta > 0$ such that $y \in \mathrm{E}$ and $\|y\| \leq \eta$ imply $\|uy\| \leq 1$. Then, if $x \in \mathrm{E}$, we have

$$\|x\| \leq 1 \Rightarrow \|\eta x\| \leq \eta \Rightarrow \|u(\eta x)\| \leq 1 \Rightarrow \|ux\| \leq \eta^{-1},$$

therefore $\|u\| < +\infty$.

(iv) ⇒ (iii). Suppose $\|u\| < +\infty$. Let $x, y \in \mathrm{E}$ and $\varepsilon > 0$. Then

$$\|x - y\| \leq \frac{\varepsilon}{\|u\|} \Rightarrow \|ux - uy\| = \|u(x - y)\| \leq \|u\|\,\|x - y\| \leq \varepsilon,$$

thus u is uniformly continuous.

8.2.5. Condition (iv) of 8.2.4 means, in the notations of 8.2.1, that $u(\mathrm{B})$ is bounded. The expression 'bounded linear mapping' is used as a synonym for 'continuous linear mapping'.

8.2.6. Example. Let $\mathrm{E} = \mathscr{C}([0, 1], \mathbf{R})$, equipped with the norm of uniform convergence. The mapping $f \mapsto f(0)$ of E into $\mathbf{R}$ is linear; it is continuous because $|f(0)| \leq \|f\|$ for all $f \in \mathrm{E}$.

Let F be the linear subspace of E formed by the differentiable functions, equipped with the norm induced by that of E. The mapping $f \mapsto f'(0)$ of F into $\mathbf{R}$ is linear; it is not continuous, since, for every number $\mathrm{A} > 0$, one can construct a function $f \in F$ such that $|f(x)| \leq 1$ for all $x \in [0, 1]$ but $f'(0) \geq A$ (for example, $f(x) = Ax/(1 + Ax)$).

8.2.7. Let E, F be normed spaces. *We denote by* $\mathscr{L}(E, F)$ the set of all continuous linear mappings of E into F. If $u, v \in \mathscr{L}(E, F)$ and λ is a scalar, then $u + v \in \mathscr{L}(E, F)$ and $\lambda u \in \mathscr{L}(E, F)$ by 8.1.8. Thus $\mathscr{L}(E, F)$ is in a natural way a real or complex vector space. When E = F, one writes $\mathscr{L}(E) = \mathscr{L}(E, E)$. The identity mapping id_E, also denoted 1 or I, is an element of $\mathscr{L}(E)$. If $u \in \mathscr{L}(E)$ and $v \in \mathscr{L}(E)$ then $u \circ v \in \mathscr{L}(E)$, so that $\mathscr{L}(E)$ is in a natural way an algebra over **R** or **C** with unity element 1.

8.2.8. Theorem. *Let* E, F, G *be normed spaces.*

(i) *The mapping* $u \mapsto \|u\|$ *of* $\mathscr{L}(E, F)$ *into* $[0, +\infty)$ *is a norm on* $\mathscr{L}(E, F)$.
(ii) *If* $u \in \mathscr{L}(E, F)$ *and* $v \in \mathscr{L}(F, G)$, *then* $\|v \circ u\| \leq \|v\| \|u\|$.
(iii) $\|id_E\| = 1$ *if* $E \neq 0$.

(i) Let $u, v \in \mathscr{L}(E, F)$. For every $x \in E$,

$$\begin{aligned}\|(u + v)(x)\| &= \|ux + vx\| \leq \|ux\| + \|vx\| \\ &\leq \|u\| \|x\| + \|v\| \|x\| = (\|u\| + \|v\|)\|x\|,\end{aligned}$$

therefore $\|u + v\| \leq \|u\| + \|v\|$. Also,

$$\begin{aligned}\|\lambda u\| &= \sup_{\|x\| \leq 1} \|(\lambda u)(x)\| = \sup_{\|x\| \leq 1} |\lambda| \|ux\| \\ &= |\lambda| \sup_{\|x\| \leq 1} \|ux\| = |\lambda| \|u\|.\end{aligned}$$

Finally, if $\|u\| = 0$ then $\|ux\| = 0$ for all $x \in E$, therefore $ux = 0$ for all $x \in E$, thus $u = 0$.

(ii) Let $u \in \mathscr{L}(E, F)$, $v \in \mathscr{L}(F, G)$. For every $x \in E$,

$$\|(v \circ u)(x)\| \leq \|v\| \|ux\| \leq \|v\| \|u\| \|x\|,$$

therefore $\|v \circ u\| \leq \|v\| \|u\|$.

(iii) This is obvious.

8.2.9. Thus, $\mathscr{L}(E, F)$ is in a natural way a normed vector space. The norm defines a metric and a topology on $\mathscr{L}(E, F)$. This topology is called the *norm topology* on $\mathscr{L}(E, F)$. By 8.1.8, the mappings

$$(u, v) \mapsto u + v \quad \text{of} \quad \mathscr{L}(E, F) \times \mathscr{L}(E, F) \quad \text{into} \quad \mathscr{L}(E, F),$$

$$(\lambda, u) \mapsto \lambda u \quad \text{of} \quad \mathbf{R} \times \mathscr{L}(E, F) \quad (\text{or } \mathbf{C} \times \mathscr{L}(E, F)) \quad \text{into} \quad \mathscr{L}(E, F),$$

are continuous. It follows easily from 8.2.8(ii) that the mapping

$$(u, v) \mapsto v \circ u \quad \text{of} \quad \mathscr{L}(E, F) \times \mathscr{L}(F, G) \quad \text{into} \quad \mathscr{L}(E, G)$$

is continuous.

8.2.10. Let E be a real or complex vector space. The elements of $\mathscr{L}(\mathrm{E}, \mathbf{R})$ or $\mathscr{L}(\mathrm{E}, \mathbf{C})$ are the *continuous linear forms* on E. The space $\mathscr{L}(\mathrm{E}, \mathbf{R})$ or $\mathscr{L}(\mathrm{E}, \mathbf{C})$ is a normed vector space, called the *dual* of E and often denoted E′.

8.3. Bicontinuous Linear Mappings

8.3.1. Theorem. *Let* E, F *be normed spaces, u a linear mapping of* E onto F. *The following conditions are equivalent:*

(i) *u is bijective and bicontinuous;*
(ii) *there exist numbers a,* $\mathrm{A} \in (0, +\infty)$ *such that*

$$a\|x\| \leq \|ux\| \leq \mathrm{A}\|x\|$$

for all $x \in \mathrm{E}$;
(iii) *there exist numbers a,* $\mathrm{A} \in (0, +\infty)$ *such that*

$$a \leq \|ux\| \leq \mathrm{A}$$

for all $x \in \mathrm{E}$ *with* $\|x\| = 1$.

(i) ⇒ (ii). Suppose u is bijective and bicontinuous. There exist $\mathrm{A}, \mathrm{B} \in (0, +\infty)$ such that $\|ux\| \leq \mathrm{A}\|x\|$ for all $x \in \mathrm{E}$ and $\|u^{-1}y\| \leq \mathrm{B}\|y\|$ for all $y \in \mathrm{F}$ (8.2.4). Let $x \in \mathrm{E}$. Set $y = ux$, so that $x = u^{-1}y$. Then $\|x\| \leq \mathrm{B}\|y\|$, that is, $(1/\mathrm{B})\|x\| \leq \|ux\|$, and $1/\mathrm{B} > 0$ since $\mathrm{B} < +\infty$.

(ii) ⇒ (i). Suppose condition (ii) is satisfied. Then u is continuous (8.2.4). Next, $ux = 0$ implies $a\|x\| = 0$, therefore $x = 0$ (because $a > 0$); this proves that u (which is surjective by hypothesis) is bijective. Finally, let $y \in \mathrm{F}$. Set $x = u^{-1}y$. Then $y = ux$, and the inequality $a\|x\| \leq \|ux\|$ may be written $a\|u^{-1}y\| \leq \|y\|$, or $\|u^{-1}y\| \leq (1/a)\|y\|$. Thus u^{-1} is continuous (8.2.4).

(ii) ⇒ (iii). This is obvious.

(iii) ⇒ (ii). Suppose that condition (iii) is satisfied and let us prove that $a\|x\| \leq \|ux\| \leq \mathrm{A}\|x\|$ for all $x \in \mathrm{E}$. This is clear if $x = 0$. If $x \neq 0$, let $x' = x/\|x\|$. Then $\|x'\| = 1$, therefore $a \leq \|ux'\| \leq \mathrm{A}$. Now, $ux = u(\|x\|x') = \|x\|ux'$, therefore $a\|x\| \leq \|ux\| \leq \mathrm{A}\|x\|$.

8.3.2. Theorem. *Let* $x \mapsto \|x\|_1$ *and* $x \mapsto \|x\|_2$ *be two norms on a vector space* E. *The following conditions are equivalent:*

(i) *the topologies defined by the two norms on* E *are the same;*
(ii) *the identity mapping of* E_1 *into* E_2 *(where* $\mathrm{E}_1, \mathrm{E}_2$ *denote the vector space* E *equipped with the norms* $x \mapsto \|x\|_1$, $x \mapsto \|x\|_2$*) is bicontinuous;*
(iii) *there exist numbers a,* $\mathrm{A} \in (0, +\infty)$ *such that*

$$a\|x\|_1 \leq \|x\|_2 \leq \mathrm{A}\|x\|_1$$

for all $x \in \mathrm{E}$.

The equivalence (i) $\Leftrightarrow$ (ii) is obvious for any pair of topologies on a set (cf. 2.4.7). The equivalence (ii) $\Leftrightarrow$ (iii) follows from 8.3.1.

8.3.3. Definition. If two norms on a vector space satisfy the conditions of 8.3.2, they are said to be *equivalent*. This is clearly an equivalence relation among norms.

8.3.4. For example, it is well known that, on $\mathbf{R}^n$ or $\mathbf{C}^n$, the three norms defined in 8.1.2 are equivalent. More generally:

▶ **Theorem.** *On* $\mathbf{R}^n$ *or* $\mathbf{C}^n$, *all norms are equivalent.*

Let us treat for example the case of $\mathbf{R}^n$. Let $x \mapsto \|x\|$ be a norm on $\mathbf{R}^n$. Let $x \mapsto \|x\|'$ be the norm $(x_1, \ldots, x_n) \mapsto |x_1| + \cdots + |x_n|$. It suffices to prove that these two norms are equivalent. In the proof, we are going to use topological concepts in $\mathbf{R}^n$; the only topology on $\mathbf{R}^n$ we shall use will be the usual topology, which is defined by the norm $x \mapsto \|x\|'$ (cf. 8.1.7, 1.1.15, 1.1.2).

Let $(e_1, \ldots, e_n)$ be the canonical basis of $\mathbf{R}^n$. Set

$$\mathrm{A} = \sup(\|e_1\|, \ldots, \|e_n\|).$$

If $x = (x_1, \ldots, x_n) \in \mathbf{R}^n$, then

$$\begin{aligned}\|x\| &= \|x_1e_1 + \cdots + x_ne_n\| \leq |x_1|\,\|e_1\| + \cdots + |x_n|\,\|e_n\| \\ &\leq \mathrm{A}(|x_1| + \cdots + |x_n|) = \mathrm{A}\|x\|'.\end{aligned}$$

It follows from this that the function $x \mapsto \|x\|$ on $\mathbf{R}^n$ is continuous; for, if $x, x_0 \in \mathbf{R}^n$ and $\varepsilon > 0$, then

$$\|x - x_0\|' \leq \frac{\varepsilon}{\mathrm{A}} \Rightarrow \|x - x_0\| \leq \varepsilon \Rightarrow \big|\|x\| - \|x_0\|\big| \leq \varepsilon.$$

Let S be the set of $x = (x_1, \ldots, x_n) \in \mathbf{R}^n$ such that

$$\|x\|' = |x_1| + \cdots + |x_n| = 1.$$

This is a closed set in $\mathbf{R}^n$ (1.1.12), clearly bounded, hence compact (4.2.18). The function $x \mapsto \|x\|$ is continuous on S by what was shown earlier, and it does not vanish on S, therefore there exists a number $a > 0$ such that $\|x\| \geq a$ for all $x \in \mathrm{S}$ (4.2.14). Since, moreover, $\|x\| \leq \mathrm{A}$ for all $x \in \mathrm{S}$, we see that the two norms are equivalent.

8.3.5. Corollary. *On a finite-dimensional real or complex vector space, all norms are equivalent.*

For, such a vector space is isomorphic to $\mathbf{R}^n$ or $\mathbf{C}^n$ for some n.

8.3.6. Thus, on a finite-dimensional real or complex vector space E, there exists a 'natural' topology $\mathscr{T}$: the topology defined by any norm on E. If u

is any isomorphism of the vector space $\mathbf{R}^n$ or $\mathbf{C}^n$ onto E, then $\mathcal{T}$ is the transport by u of the usual topology of $\mathbf{R}^n$ or $\mathbf{C}^n$.

8.3.7. Let E, F be normed vector spaces, u a linear mapping of E into F. If E is finite-dimensional, then u is automatically *continuous*. For, we can suppose by 8.3.6 that $E = \mathbf{R}^n$ (or $\mathbf{C}^n$). Let $(e_1, \ldots, e_n)$ be the canonical basis of $\mathbf{R}^n$ (for example). Let $a_i = u(e_i) \in F$. For every $x = (x_1, \ldots, x_n) \in \mathbf{R}^n$, we have $u(x) = x_1 a_1 + \cdots + x_n a_n$. The mapping $x \mapsto x_1$ of $\mathbf{R}^n$ into $\mathbf{R}$ is continuous (3.2.8). The mapping $\lambda \mapsto \lambda a_1$ of $\mathbf{R}$ into F is continuous (8.1.8). Therefore the mapping $x \mapsto x_1 a_1$ of $\mathbf{R}^n$ into F is continuous. Similarly for the mappings $x \mapsto x_2 a_2, \ldots, x \mapsto x_n a_n$. Therefore the mapping $x \mapsto (x_1 a_1, \ldots, x_n a_n)$ of $\mathbf{R}^n$ into F^n is continuous (3.2.7). Consequently, the mapping $x \mapsto x_1 a_1 + \cdots + x_n a_n$ of $\mathbf{R}^n$ into F is continuous (8.1.8).

However, if E is infinite-dimensional then u may be discontinuous, as we saw in 8.2.6.

8.3.8. If E and F are normed spaces of finite dimensions m and n, then, by 8.3.7, $\mathscr{L}(E, F)$ is the vector space of *all* linear mappings of E into F. This vector space has dimension mn, so by 8.3.6 it possesses a natural topology $\mathcal{T}$. If one chooses bases in E and F, there is a canonical linear bijection $u \mapsto M_u$ of $\mathscr{L}(E, F)$ onto the vector space $M_{n,m}$ of real or complex matrices with n rows and m columns. Under this bijection, the topology $\mathcal{T}$ corresponds to the natural topology of $M_{n,m}$; the latter topology is defined, for example, by the norm $(\alpha_{ij})_{1 \le i \le n,\, 1 \le j \le m} \mapsto \sum_{i,j} |\alpha_{ij}|$.

8.4. Pre-Hilbert Spaces

8.4.1. Definition. A *complex pre-Hilbert space* is a complex vector space E equipped with a mapping $(x, y) \mapsto (x|y)$ of $E \times E$ into $\mathbf{C}$, called the *scalar product* (or 'inner product'), satisfying the following conditions:

(i) $(x|y)$ depends linearly on y, for fixed x;
(ii) $(x|y) = \overline{(y|x)}$ for $x, y \in E$ (so that $(x|y)$ depends 'conjugate-linearly' on x, for fixed y);
(iii) $(x|x) \ge 0$ for $x \in E$.

There is an obvious notion of isomorphism of pre-Hilbert spaces.

8.4.2. Example. For

$$x = (x_1, \ldots, x_n) \in \mathbf{C}^n \quad \text{and} \quad y = (y_1, \ldots, y_n) \in \mathbf{C}^n,$$

set

$$(x|y) = \bar{x}_1 y_1 + \bar{x}_2 y_2 + \cdots + \bar{x}_n y_n.$$

One verifies immediately that $(x, y) \mapsto (x|y)$ is a scalar product on $\mathbf{C}^n$, called the *canonical scalar product.*

8.4.3. Example. Let E be the complex vector space whose elements are the sequences $(\lambda_1, \lambda_2, \ldots)$ of complex numbers that are zero from some index onward. For

$$x = (\lambda_1, \lambda_2, \ldots) \in \mathrm{E} \quad \text{and} \quad y = (\mu_1, \mu_2, \ldots) \in \mathrm{E}$$

set

$$(x|y) = \bar{\lambda}_1 \mu_1 + \bar{\lambda}_2 \mu_2 + \cdots$$

(this sum involves only a finite number of nonzero terms). One verifies immediately that $(x, y) \mapsto (x|y)$ is a scalar product on E.

8.4.4. Example. More generally, let I be a set. Let E be the complex vector space whose elements are the families $(\lambda_i)_{i \in \mathrm{I}}$ of complex numbers such that $\lambda_i = 0$ for almost all $i \in \mathrm{I}$ (cf. 3.3.1). For $x = (\lambda_i)_{i \in \mathrm{I}} \in \mathrm{E}$ and $y = (\mu_i)_{i \in \mathrm{I}} \in \mathrm{E}$, set

$$(x|y) = \sum_{i \in \mathrm{I}} \bar{\lambda}_i \mu_i.$$

This is a scalar product on E. If $\mathrm{I} = \{1, 2, \ldots, n\}$, one recovers Example 8.4.2. If $\mathrm{I} = \{1, 2, 3, \ldots\}$, one recovers Example 8.4.3.

8.4.5. Example. Let E be the set of continuous complex-valued functions on $[0, 1]$. For $g, h \in \mathrm{E}$, set

$$(g|h) = \int_0^1 \overline{g(t)} h(t)\, dt.$$

This is a scalar product on E.

8.4.6. Example. Let X be the vector space of all sequences $(\lambda_1, \lambda_2, \ldots)$ of complex numbers. We denote by $l^2_{\mathbf{C}}$, or simply l^2, the set of sequences $(\lambda_1, \lambda_2, \ldots) \in \mathrm{X}$ such that $\sum_{n=1}^{\infty} |\lambda_n|^2 < +\infty$. Let us show that *this is a linear subspace of* X. It is clear that if $s \in l^2$ and $\lambda \in \mathbf{C}$, then $\lambda s \in l^2$. Let

$$s = (\lambda_1, \lambda_2, \ldots) \in l^2 \quad \text{and} \quad t = (\mu_1, \mu_2, \ldots) \in l^2.$$

Recall that if $\alpha, \beta \in \mathbf{C}$ then

$$\begin{aligned} |\alpha + \beta|^2 + |\alpha - \beta|^2 &= (\bar{\alpha} + \bar{\beta})(\alpha + \beta) + (\bar{\alpha} - \bar{\beta})(\alpha - \beta) \\ &= 2\bar{\alpha}\alpha + 2\bar{\beta}\beta = 2|\alpha|^2 + 2|\beta|^2; \end{aligned}$$

therefore $|\alpha + \beta|^2 \leq 2|\alpha|^2 + 2|\beta|^2$. This established, we have

$$\sum_{n=1}^{\infty} |\lambda_n + \mu_n|^2 \leq \sum_{n=1}^{\infty} (2|\lambda_n|^2 + 2|\mu_n|^2)$$
$$= 2\sum_{n=1}^{\infty} |\lambda_n|^2 + 2\sum_{n=1}^{\infty} |\mu_n|^2 < +\infty,$$

thus $s + t \in l^2$.

Moreover, $0 \leq (|\alpha| - |\beta|)^2 = |\alpha|^2 + |\beta|^2 - 2|\alpha||\beta|$, therefore

$$2\sum_{n=1}^{\infty} |\lambda_n||\mu_n| \leq \sum_{n=1}^{\infty} (|\lambda_n|^2 + |\mu_n|^2) < +\infty,$$

so that the series $\sum_{n=1}^{\infty} \bar{\lambda}_n \mu_n$ is absolutely convergent. Set

$$(s|t) = \bar{\lambda}_1\mu_1 + \bar{\lambda}_2\mu_2 + \cdots.$$

We obtain in this way a scalar product on l^2.

8.4.7. Let E be a pre-Hilbert space, E′ a linear subspace of E. The scalar product of E, restricted to E′, is a scalar product on E′. Thus E′ automatically becomes a pre-Hilbert space.

For example, the pre-Hilbert space of 8.4.3 is a pre-Hilbert subspace of l^2.

8.4.8. Let E be a pre-Hilbert space, and $x, y \in$ E. We say that x, y are *orthogonal* if $(x|y) = 0$. This relation between x and y is symmetric. We say that subsets M, N of E are orthogonal if every element of M is orthogonal to every element of N. If M is orthogonal to N, then every linear combination of elements of M is orthogonal to every linear combination of elements of N.

8.4.9. Let M $\subset$ E. The set of elements of E orthogonal to M is a linear subspace of E that is denoted $\mathrm{M}^{\perp}$ and is called, through misuse of language, *the* linear subspace of E orthogonal to M.

8.4.10. Theorem (Cauchy–Schwarz Inequality). *Let* E *be a pre-Hilbert space. For all* $x, y \in$ E,

$$|(x|y)|^2 \leq (x|x)(y|y).$$

For all $\lambda \in \mathbf{C}$, we have

$$(1)\qquad 0 \leq (x + \lambda y|x + \lambda y)$$
$$= \bar{\lambda}\lambda(y|y) + \bar{\lambda}(y|x) + \lambda(x|y) + (x|x).$$

Multiply through by $(y|y)$; after calculation, we obtain

$$(2)\quad 0 \leq [\bar{\lambda}(y|y) + (x|y)][\lambda(y|y) + (y|x)] + (x|x)(y|y) - (x|y)(y|x).$$

Suppose first that $(y|y) \neq 0$. We can then set

$$\lambda = -(y|x)/(y|y)$$

in (2), obtaining

$$0 \leq (x|x)(y|y) - (x|y)(y|x),$$

which is the inequality of the theorem. If $(x|x) \neq 0$, we need only interchange the roles of x and y in the preceding. Finally, if $(x|x) = (y|y) = 0$ then (1) reduces to

$$0 \leq \bar{\lambda}(y|x) + \lambda(x|y). \tag{3}$$

Taking $\lambda = -(y|x)$ in (3), we get

$$0 \leq -(x|y)\overline{(x|y)} - \overline{(x|y)}(x|y) = -2|(x|y)|^2,$$

therefore $(x|y) = 0$ and the inequality of the theorem is again verified.

8.5. Separated Pre-Hilbert Spaces

8.5.1. Theorem. *Let* E *be a pre-Hilbert space, and* x *an element of* E. *The following conditions are equivalent*:

(i) $x \in E^{\perp}$ *(in other words,* $(x|y) = 0$ *for all* $y \in E$);
(ii) $(x|x) = 0$.

(i) ⇒ (ii). This is obvious.
(ii) ⇒ (i). Suppose $(x|x) = 0$. By 8.4.10, for every $y \in E$ we have $|(x|y)|^2 = 0$, therefore $(x|y) = 0$.

8.5.2. The linear subspace $E^{\perp}$ of E is called the *kernel* of the scalar product. If $E^{\perp} = 0$, we say that the scalar product is *non-degenerate*, or that the pre-Hilbert space is *separated* (cf. 8.5.8). By 8.5.1, this amounts to saying that $(x|x) > 0$ for every nonzero vector x of E.

8.5.3. One verifies easily that the pre-Hilbert spaces defined in 8.4.2–8.4.6 are separated. However, if E is a nonzero complex vector space and if one sets $(x|y) = 0$ for all $x, y \in E$, then the pre-Hilbert space so obtained is not separated.

8.5.4. Let E be a pre-Hilbert space, $E^{\perp}$ its kernel, E′ the complex vector space $E/E^{\perp}$. We are going to define a scalar product on E′. Let $x', y' \in E'$. Choose representatives x, y of x', y' in E. Then the number $(x|y)$ depends only on x', y' and not on the choice of representatives. For, any other representatives are of the form $x_1 = x + u$, $y_1 = y + v$ with $u, v \in E^{\perp}$, whence

$$\begin{aligned}(x_1|y_1) &= (x|y) + (x|v) + (u|y) + (u|v)\\ &= (x|y) + 0 + 0 + 0 = (x|y).\end{aligned}$$

We can therefore define $(x'|y') = (x|y)$. The fact that $(x, y) \mapsto (x|y)$ is a scalar product on E implies easily that $(x', y') \mapsto (x'|y')$ is a scalar product on E'. Thus E' is a pre-Hilbert space. Let us show that E' is *separated*. Let x' be a nonzero element of E'. Let x be a representative of x' in E. Then $x \notin E^{\perp}$ (otherwise, $x' = 0$), therefore $(x|x) > 0$ and consequently $(x'|x') > 0$. We say that E' is *the separated pre-Hilbert space associated with* E. This construction reduces most problems about pre-Hilbert spaces to problems about separated pre-Hilbert spaces.

8.5.5. Theorem. *Let* E *be a pre-Hilbert space. For every* $x \in E$, *set* $\|x\| = \sqrt{(x|x)}$. *Then* $x \mapsto \|x\|$ *is a seminorm on* E. *For this seminorm to be a norm, it is necessary and sufficient that the pre-Hilbert space be separated.*

If $x, y \in E$ and $\lambda \in \mathbf{C}$, then

$$\begin{aligned}\|\lambda x\|^2 &= (\lambda x|\lambda x) = \bar{\lambda}\lambda(x|x) = |\lambda|^2\|x\|^2,\\ \|x + y\|^2 &= (x + y|x + y) = (x|x) + (y|y) + 2\,\mathrm{Re}(x|y)\\ &\le \|x\|^2 + \|y\|^2 + 2|(x|y)|\\ &\le \|x\|^2 + \|y\|^2 + 2\|x\|\,\|y\| \quad \text{by 8.4.10}\\ &= (\|x\| + \|y\|)^2,\end{aligned}$$

therefore $x \mapsto \|x\|$ is a seminorm. For this seminorm to be a norm, it is necessary and sufficient that

$$x \neq 0 \Rightarrow \|x\| > 0,$$

in other words that

$$x \neq 0 \Rightarrow (x|x) > 0.$$

8.5.6. Theorem. *Let* E *be a pre-Hilbert space and* $x, y \in E$. *Then*:

(i) $\|x + y\|^2 + \|x - y\|^2 = 2\|x\|^2 + 2\|y\|^2$ (*parallelogram law*),

and

(ii) $4(x|y) = \|x + y\|^2 - \|-x + y\|^2 + i\|ix + y\|^2 - i\|-ix + y\|^2$ (*polarization identity*).

For

$$\begin{aligned}\|x + y\|^2 + \|x - y\|^2 &= (x + y|x + y) + (x - y\ x - y)\\ &= (x|x) + (x|y) + (y|x) + (y|y)\\ &\quad + (x|x) - (x|y) - (y|x) + (y|y)\\ &= 2(x|x) + 2(y|y),\end{aligned}$$

and

$$\begin{aligned} (x+y|x+y) &= (x|x)+(x|y)+(y|x)+(y|y), \\ -(-x+y|-x+y) &= -(x|x)+(x|y)+(y|x)-(y|y), \\ i(ix+y|ix+y) &= i(x|x)+(x|y)-(y|x)+i(y|y), \\ -i(-ix+y|-ix+y) &= -i(x|x)+(x|y)-(y|x)-i(y|y). \end{aligned}$$

8.5.7. Theorem (Pythagoras). *Let* E *be a pre-Hilbert space, and* $x_1, \ldots, x_n$ *pairwise orthogonal elements of* E. *Then*

$$\|x_1 + \cdots + x_n\|^2 = \|x_1\|^2 + \cdots + \|x_n\|^2.$$

For,

$$\left(\sum_{i=1}^{n} x_i \,\middle|\, \sum_{j=1}^{n} x_j\right) = \sum_{i,j=1}^{n} (x_i|x_j) = \sum_{i=1}^{n} (x_i|x_i).$$

8.5.8. Let E be a separated pre-Hilbert space. By 8.5.5, E has a norm $x \mapsto \|x\|$. Therefore, by 8.1.6, E has a metric

$$d(x, y) = \|x - y\| = (x - y|x - y)^{1/2},$$

hence a topology. This topology is separated (1.6.2(a)), which in some measure justifies the expression 'separated pre-Hilbert space'.

8.5.9. Examples. If $\mathbf{C}^n$ is equipped with the canonical scalar product, one recovers the norm and metric already considered. In l^2, one has

$$d((\lambda_1, \lambda_2, \ldots), (\mu_1, \mu_2, \ldots)) = (|\lambda_1 - \mu_1|^2 + |\lambda_2 - \mu_2|^2 + \cdots)^{1/2}.$$

In the pre-Hilbert space $\mathscr{C}([0, 1], \mathbf{C})$ of 8.4.5, one has

$$d(f, g) = \left(\int_0^1 |f(t) - g(t)|^2 \, dt\right)^{1/2};$$

the corresponding topology is called *the topology of convergence in mean square.* If a sequence (f_n) tends to f for this topology, one says that (f_n) tends to f in mean square.

8.5.10. Theorem. *Let* E *be a separated pre-Hilbert space. The mapping* $(x, y) \mapsto (x|y)$ *of* E × E *into* **C** *is continuous.*

The mapping $x \mapsto \|x\| = d(x, 0)$ is continuous (5.1.1). Moreover, the mappings $(x, y) \mapsto x + y$, $(\lambda, x) \mapsto \lambda x$ are continuous (8.1.8). The theorem then follows from the polarization identity (8.5.6).

8.6. Banach Spaces, Hilbert Spaces

8.6.1. Definition. A normed space that is complete (as a metric space) is called a *Banach space.* A complete, separated pre-Hilbert space is called a *Hilbert space.*

8.6.2. Example. Let E be a finite-dimensional normed space. Then E is a Banach space. For, one can suppose that $E = \mathbf{R}^n$ or $\mathbf{C}^n$. The given norm is equivalent to the usual norm of $\mathbf{R}^n$ or $\mathbf{C}^n$ (8.3.4). Every Cauchy sequence for the given norm is therefore a Cauchy sequence for the usual norm, hence has a limit (5.5.9). Thus E is complete.

In particular, a finite-dimensional separated pre-Hilbert space (for example, $\mathbf{C}^n$ equipped with the canonical scalar product) is a Hilbert space.

8.6.3. Example. Let X be a set, E the vector space of bounded real-valued functions on X, equipped with the norm of uniform convergence. Then E is a Banach space. For, $\mathscr{F}(X, \mathbf{R})$ is complete for the metric of uniform convergence (6.1.6). If a sequence (f_n) of functions in E tends uniformly to a function $f \in \mathscr{F}(X, \mathbf{R})$, then $\sup_{x \in X} |f_n(x) - f(x)| \leq 1$ for n sufficiently large, therefore f is bounded. Thus E is closed in $\mathscr{F}(X, \mathbf{R})$ hence is complete.

In particular, l^∞ is a Banach space.

8.6.4. Example. Let us show that l^2 is complete (hence is a Hilbert space). For $n = 1, 2, \ldots$, let $x_n \in l^2$. Then $x_n = (\lambda_{n1}, \lambda_{n2}, \lambda_{n3}, \ldots)$ with $\sum_{i=1}^{\infty} |\lambda_{ni}|^2 < +\infty$. Suppose that $\|x_p - x_q\| \to 0$ as $p, q \to \infty$, and let us show that (x_n) tends to an element of l^2. We have $\sum_{i=1}^{\infty} |\lambda_{pi} - \lambda_{qi}|^2 \to 0$ as $p, q \to \infty$. *A fortiori*, for every fixed positive integer i, $|\lambda_{pi} - \lambda_{qi}| \to 0$ as $p, q \to \infty$, therefore (λ_{ni}) tends to a complex number λ_i as $n \to \infty$. Let $\varepsilon > 0$. There exists a positive integer N such that

$$p, q \geq \mathrm{N} \Rightarrow \sum_{i=1}^{\infty} |\lambda_{pi} - \lambda_{qi}|^2 \leq \varepsilon.$$

Let A be a positive integer, provisionally fixed. We have

$$p, q \geq \mathrm{N} \Rightarrow \sum_{i=1}^{\mathrm{A}} |\lambda_{pi} - \lambda_{qi}|^2 \leq \varepsilon.$$

Fix $p \geq \mathrm{N}$ and let $q \to \infty$. We obtain:

$$p \geq \mathrm{N} \Rightarrow \sum_{i=1}^{\mathrm{A}} |\lambda_{pi} - \lambda_i|^2 \leq \varepsilon.$$

This inequality being true for every positive integer A, we deduce that

$$(1) \qquad p \geq \mathrm{N} \Rightarrow \sum_{i=1}^{\infty} |\lambda_{pi} - \lambda_i|^2 \leq \varepsilon.$$

This proves, first of all, that for $p \geq \mathrm{N}$ the sequence $(\lambda_{pi} - \lambda_i)$ belongs to l^2. Since $\lambda_i = \lambda_{pi} - (\lambda_{pi} - \lambda_i)$, we deduce that (λ_i) is an element x of l^2. The relation (1) can now be written

$$p \geq \mathrm{N} \Rightarrow \|x_p - x\|^2 \leq \varepsilon.$$

Thus $x_p \to x$ as $p \to \infty$.

8.6.5. Example. An almost identical proof shows that l^1 is a Banach space.

8.6.6. Example. The separated pre-Hilbert space of 8.4.3 is not complete (cf. 8.7.1).

8.6.7. Theorem. (i) *A closed linear subspace of a Banach space is a Banach space.*

(ii) *A finite product of Banach spaces is a Banach space.*

Assertion (i) follows from 5.5.6, assertion (ii) from 5.5.8.

8.6.8. Theorem. *Let* E *be a normed space,* F *a Banach space. Then* $\mathscr{L}(\mathrm{E}, \mathrm{F})$ *is a Banach space.*

Let (f_n) be a Cauchy sequence in $\mathscr{L}(\mathrm{E}, \mathrm{F})$. Let $x \in \mathrm{E}$. We have

$$\|f_m(x) - f_n(x)\| \leq \|f_m - f_n\| \|x\| \to 0$$

as $m, n \to \infty$. Since F is complete, there exists an element $f(x)$ of F such that $f_n(x) \to f(x)$ as $n \to \infty$. We have thus defined a mapping f of E into F. The equality $f_n(x + y) = f_n(x) + f_n(y)$, valid for all n, yields $f(x + y) = f(x) + f(y)$ in the limit; similarly $f(\lambda x) = \lambda f(x)$ for every scalar λ. Thus f is linear. Let $\varepsilon > 0$. There exists an N such that $m, n \geq \mathrm{N} \Rightarrow \|f_m - f_n\| \leq \varepsilon$, that is, $\|f_m(x) - f_n(x)\| \leq \varepsilon$ for all $x \in \mathrm{E}$ such that $\|x\| \leq 1$. Fixing x and m, and letting n tend to infinity, we obtain $\|f_m(x) - f(x)\| \leq \varepsilon$ for $\|x\| \leq 1$. From this, we deduce first of all that

$$\|f(x)\| \leq \varepsilon + \|f_m(x)\| \leq \varepsilon + \|f_m\| \quad \text{for} \quad \|x\| \leq 1,$$

therefore $f \in \mathscr{L}(\mathrm{E}, \mathrm{F})$. Moreover, we see that $\|f_m - f\| \leq \varepsilon$, and this for all $m \geq \mathrm{N}$. Thus (f_n) tends to f in $\mathscr{L}(\mathrm{E}, \mathrm{F})$.

8.6.9. Corollary. *Let* E *be a normed space. Its dual is a Banach space.*

For, the dual is either $\mathscr{L}(\mathrm{E}, \mathbf{R})$ or $\mathscr{L}(\mathrm{E}, \mathbf{C})$, and **R**, **C** are complete.

* ▶ **8.6.10. Theorem** (Banach–Steinhaus). *Let* E *be a Banach space,* F *a normed space,* $(u_i)_{i \in \mathrm{I}}$ *a family of continuous linear mappings of* E *into* F. *The following conditions are equivalent:*

(i) $\sup_{i \in \mathrm{I}} \|u_i\| < +\infty$;
(ii) *for each* $x \in \mathrm{E}$, $\sup_{i \in \mathrm{I}} \|u_i x\| < +\infty$.

(i) ⇒ (ii). This is immediate since $\|u_i x\| \le \|u_i\| \, \|x\|$.

Suppose that condition (ii) is satisfied. The functions $x \mapsto \|u_i x\|$ on E are continuous, and their upper envelope is finite. By 7.4.15, there exist a closed ball B in E with center a and radius $\rho > 0$, and a constant $\mathrm{M} > 0$, such that $\|u_i x\| \le \mathrm{M}$ for all $x \in \mathrm{B}$ and $i \in \mathrm{I}$. Since $\|u_i(x - a)\| \le \|u_i x\| + \|u_i a\|$, there exists a constant $\mathrm{M}' > 0$ such that $\|u_i y\| \le \mathrm{M}'$ for $\|y\| \le \rho$ and for all $i \in \mathrm{I}$. Then, for $\|y\| \le 1$, we have

$$\|u_i y\| = \rho^{-1} \|u_i(\rho y)\| \le \rho^{-1}\mathrm{M}',$$

therefore $\|u_i\| \le \rho^{-1}\mathrm{M}'$, and this for every $i \in \mathrm{I}$.

* **8.6.11. Corollary.** *Let* E *be a Banach space,* F *a normed space, and* $(u_1, u_2, \ldots)$ *a sequence of continuous linear mappings of* E *into* F *such that, for every* $x \in \mathrm{E}$, $u_n x$ *has a limit* ux *in* F. *Then* u *is a continuous linear mapping of* E *into* F.

The fact that u is linear is immediate. By 8.6.10, there exists a finite constant M such that $\|u_n\| \le \mathrm{M}$ for all n. For every $x \in \mathrm{E}$, we have $\|u_n x\| \le \mathrm{M}\|x\|$ for all n, whence $\|ux\| \le \mathrm{M}\|x\|$ on passage to the limit. Therefore u is continuous.

8.7. Linear Subspaces of a Normed Space

8.7.1. Let X be a normed space, E a linear subspace of X. *If* E *is finite-dimensional, then* E *is closed in* X (8.6.2 and 5.5.7). In particular, in a finite-dimensional normed space, all linear subspaces are closed.

This is not so in general. For example, consider again the example E of 8.4.3, which is a linear subspace of l^2. Let us show that E is *dense* in l^2 (hence not closed, since it is distinct from l^2). Let $(\lambda_n) \in l^2$ and $\varepsilon > 0$. There exists a positive integer N such that $\sum_{n \geq N} |\lambda_n|^2 \leq \varepsilon$. Let (μ_n) be the element of E such that $\mu_n = \lambda_n$ for $n < N$, $\mu_n = 0$ for $n \geq N$. Then

$$\|(\lambda_n) - (\mu_n)\|^2 = \sum_{n=1}^{\infty} |\lambda_n - \mu_n|^2 = \sum_{n \geq N} |\lambda_n|^2 \leq \varepsilon,$$

which proves our assertion.

In particular, E is not complete.

8.7.2. Theorem. *Let* E *be a normed space,* F *a linear subspace of* E. *Then the closure* $\overline{F}$ *of* F *in* E *is a linear subspace of* E.

Let $x, y \in \overline{F}$. There exist sequences (x_n), (y_n) in F such that $x_n \to x$, $y_n \to y$. Then $x_n + y_n \in F$ and $x_n + y_n \to x + y$ (8.1.8), therefore $x + y \in \overline{F}$. If λ is a scalar, then $\lambda x_n \to \lambda x$ and $\lambda x_n \in F$, therefore $\lambda x \in \overline{F}$.

8.7.3. Theorem. *Let* E *be a normed space, and* $A \subset E$. *Let* B *be the set of linear combinations of elements of* A *and let* $C = \overline{B}$. *Then* C *is the smallest closed linear subspace of* E *containing* A.

One knows that B is a linear subspace of E, therefore C is a linear subspace of E (8.7.2), and it is clear that $C \supset A$. If C′ is a closed linear subspace of E containing A, then $B \subset C'$ since C′ is a linear subspace, therefore $C \subset C'$ since C′ is closed.

8.7.4. Definition. With the notations of 8.7.3, we say that C is *the closed linear subspace of* E *generated by* A. If C = E, we say that A is *total* in E.

8.7.5. Theorem. *Let* E, F *be normed spaces, and* $u \in \mathscr{L}(E, F)$. *Then the kernel of* u *is a closed linear subspace of* E.

For, the kernel is $u^{-1}(0)$, and it suffices to apply 2.4.4.

8.7.6. However, the range of u is in general not closed in F.

8.7.7. Theorem. *Let* E *be a normed space,* E′ *a dense linear subspace of* E, F *a Banach space, and* $u' \in \mathscr{L}(E', F)$. *There exists one and only one* $u \in \mathscr{L}(E, F)$ *that extends* u'. *One has* $\|u\| = \|u'\|$.

The uniqueness of u follows from 3.2.15.

By 8.2.4, u' is uniformly continuous. By 5.5.13, there exists a continuous mapping u of E into F that extends u'. If $x, y \in E$, there exist sequences (x_n), (y_n) in E′ such that $x_n \to x$, $y_n \to y$. Then $u'(x_n + y_n) = u'x_n + u'y_n$, which gives $u(x + y) = ux + uy$ in the limit. One sees similarly that $u(\lambda x) = \lambda ux$ for every scalar λ. Thus u is linear. Since u extends u', it is clear that $\|u\| \geq \|u'\|$. On the other hand, the inequality

$$\|ux\| \leq \|u'\|\,\|x\|$$

is true for every $x \in E'$, hence remains true for every $x \in E$ by passage to the limit, therefore $\|u\| \leq \|u'\|$.

8.7.8. Theorem. *Let* E *be a separated pre-Hilbert space, and* $A \subset E$. *Then* $A^{\perp}$ *is closed linear subspace of* E.

We already know that $A^{\perp}$ is a linear subspace of E. On the other hand, for every $x \in E$ let F_x be the set of y such that $(x \mid y) = 0$. Then F_x is closed by 2.4.4 and 8.5.10. Now, $A^{\perp} = \bigcap_{x \in A} F_x$, therefore $A^{\perp}$ is closed.

8.8. Riesz's Theorem

*► **8.8.1. Theorem** (Riesz). *Let* E *be a separated pre-Hilbert space,* F *a complete linear subspace of* E, $x \in E$, *and* δ *the distance from* x *to* F.

(i) *There exists one and only one* $y \in F$ *such that* $\|x - y\| = \delta$.
(ii) y *is the only element of* F *such that* $x - y \in F^{\perp}$.

(a) There exists a sequence (y_n) in F such that $\|x - y_n\| \to \delta$. Let $\varepsilon > 0$. There exists a positive integer N such that

$$n \geq N \Rightarrow \|x - y_n\|^2 \leq \delta^2 + \varepsilon.$$

For $m, n \geq N$, we then have

$$2\|x - y_m\|^2 + 2\|x - y_n\|^2 \leq 4\delta^2 + 4\varepsilon.$$

Applying the parallelogram law (8.5.6) to the left side, we obtain

$$\|2x - y_m - y_n\|^2 + \|y_m - y_n\|^2 \leq 4\delta^2 + 4\varepsilon,$$

or

$$\|y_m - y_n\|^2 \leq 4\delta^2 + 4\varepsilon - 4\left\|x - \frac{y_m + y_n}{2}\right\|^2.$$

Now, $\frac{1}{2}(y_m + y_n) \in F$, therefore

$$4\left\|x - \frac{y_m + y_n}{2}\right\|^2 \geq 4\delta^2,$$

so that

$$\|y_m - y_n\|^2 \leq 4\varepsilon.$$

Since F is complete, the sequence (y_n) tends to an element y of F. Since $\|x - y_n\| \to \|x - y\|$, we have $\|x - y\| = \delta$.

(b) Let $z \in F$. For all $\lambda \in \mathbf{R}$, we have

$$\|x - y\|^2 \leq \|x - (y + \lambda z)\|^2 = \|x - y\|^2 + \lambda^2\|z\|^2 - 2\,\mathrm{Re}(x - y \mid \lambda z)$$

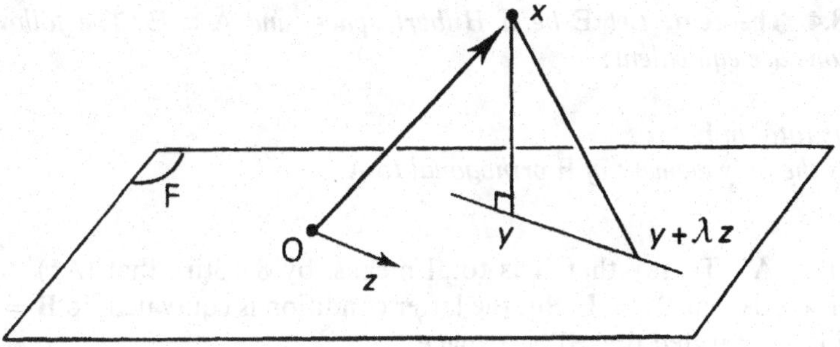

whence

$$0 \leq \lambda^2\|z\|^2 - 2\lambda\,\mathrm{Re}(x - y \mid z).$$

This requires that $\mathrm{Re}(x - y \mid z) = 0$. Replacing z by iz, we see that $(x - y \mid z) = 0$. In other words, $x - y \in F^\perp$.

(c) Let y' be an element of F distinct from y. Then $x - y$ is orthogonal to $y - y'$, since $y - y' \in F$. By the theorem of Pythagoras (8.5.7), we have

$$\|x - y'\|^2 = \|x - y\|^2 + \|y - y'\|^2 > \|x - y\|^2 = \delta^2.$$

This proves the uniqueness assertion in (i). Moreover, $x - y'$ is not orthogonal to F, since

$$\begin{aligned}(x - y' \mid y - y') &= (x - y \mid y - y') + (y - y' \mid y - y')\\ &= \|y - y'\|^2 > 0.\end{aligned}$$

This proves the uniqueness assertion in (ii).

* **8.8.2.** With the notations of 8.8.1, we say that y is the *orthogonal projection* of x on F.

* ▶ **8.8.3. Theorem.** *Let* E *be a Hilbert space.*

(i) *If* F *is a closed linear subspace of* E, *then* $(F^\perp)^\perp = F$.

(ii) *More generally, if* $A \subset E$ *then* $(A^\perp)^\perp$ *is the closed linear subspace generated by* A.

(i) It is clear that $F \subset (F^\perp)^\perp$. Let x be an element of E that does not belong to F. Let y be its orthogonal projection on F. Then $x - y \in F^\perp$, therefore $(y \mid x - y) = 0$, consequently

$$(x \mid x - y) = (x - y \mid x - y) > 0 \quad \text{because} \quad x \neq y.$$

Thus $x \notin (F^\perp)^\perp$.

(ii) Let B be the set of linear combinations of elements of A and let $C = \overline{B}$ (cf. 8.7.3 and 8.7.4). It is clear that $B^{\perp} = A^{\perp}$. We have $C^{\perp} = B^{\perp}$ on account of 8.5.10. Therefore $(A^{\perp})^{\perp} = (C^{\perp})^{\perp}$. But $(C^{\perp})^{\perp} = C$ by (i), thus $(A^{\perp})^{\perp} = C$.

*▶ **8.8.4. Theorem.** *Let* E *be a Hilbert space, and* $A \subset E$. *The following conditions are equivalent*:

(i) A *is total in* E;
(ii) 0 *is the only element of* E *orthogonal to* A.

Set $B = A^{\perp}$. To say that A is total means, by 8.8.3(ii), that $(A^{\perp})^{\perp} = E$, in other words that $B^{\perp} = E$. But the latter condition is equivalent to $B = \{0\}$ since E is a separated pre-Hilbert space.

*▶ **8.8.5. Theorem.** *Let* E *be a Hilbert space.*

(i) *For every* $x \in E$, *the mapping* $y \mapsto (x \mid y)$ *of* E *into* **C** *is a continuous linear form* f_x *on* E.
(ii) *The mapping* $x \mapsto f_x$ *of* E *into* E′ *is bijective and conjugate-linear. One has* $\|f_x\| = \|x\|$ *for all* $x \in E$.

It is clear that f_x is a linear form on E. We have

$$|f_x(y)| = |(x \mid y)| \leq \|x\| \|y\|,$$

therefore f_x is continuous and $\|f_x\| \leq \|x\|$. Let us show that $\|f_x\| = \|x\|$ for all $x \in E$. This is obvious if $x = 0$. Suppose $x \neq 0$; then $\|x\|^2 = f_x(x) \leq \|f_x\| \|x\|$, thus $\|x\| \leq \|f_x\|$ after cancelling $\|x\|$.

Let $x_1, x_2 \in E$ and $\lambda_1, \lambda_2 \in \mathbf{C}$. For all $y \in E$,

$$\begin{aligned} f_{\lambda_1 x_1 + \lambda_2 x_2}(y) &= (\lambda_1 x_1 + \lambda_2 x_2 \mid y) \\ &= \overline{\lambda}_1 (x_1 \mid y) + \overline{\lambda}_2 (x_2 \mid y) \\ &= \overline{\lambda}_1 f_{x_1}(y) + \overline{\lambda}_2 f_{x_2}(y) \\ &= (\overline{\lambda}_1 f_{x_1} + \overline{\lambda}_2 f_{x_2})(y), \end{aligned}$$

thus $f_{\lambda_1 x_1 + \lambda_2 x_2} = \overline{\lambda}_1 f_{x_1} + \overline{\lambda}_2 f_{x_2}$. In other words, the mapping $x \mapsto f_x$ of E into E′ is conjugate-linear. In view of the equality $\|f_x\| = \|x\|$, it is injective.

Let us show that it is surjective. Let $f \in E'$ and let us prove that there exists an $x \in E$ such that $f = f_x$. This is obvious if $f = 0$. Assume $f \neq 0$. Then the kernel F of f is a closed linear subspace of E distinct from E. Therefore $F^{\perp} \neq 0$ (8.8.3). Choose a nonzero element t of $F^{\perp}$. Then $t \notin F$ (otherwise $(t \mid t) = 0$), thus $f(t) \neq 0$. Multiplying t by a suitable scalar, we can suppose that $f(t) = 1$. Let $x = \|t\|^{-2} t$. For all $y \in E$,

$$f(y - f(y)t) = f(y) - f(y)f(t) = 0,$$

thus $y - f(y)t \in \mathrm{F}$. Consequently

$$f_x(y) = (x\,|\,y) = (x\,|\,f(y)t) = f(y)(x\,|\,t) = f(y)\|t\|^{-2}(t\,|\,t) = f(y).$$

8.8.6. Let E be a separated pre-Hilbert space. One calls *orthonormal family* in E a family $(e_i)_{i\in \mathrm{I}}$ of elements of E such that $\|e_i\| = 1$ for all $i \in \mathrm{I}$ and $(e_i\,|\,e_j) = 0$ for $i \in \mathrm{I}$, $j \in \mathrm{J}$, $i \neq j$. Such a family is *linearly independent*, because if

$$\lambda_{i_1}e_{i_1} + \cdots + \lambda_{i_n}e_{i_n} = 0$$

for scalars $\lambda_{i_1}, \ldots, \lambda_{i_n}$, then the Pythagorean theorem implies that

$$0 = \|\lambda_{i_1}e_{i_1} + \cdots + \lambda_{i_n}e_{i_n}\|^2 = |\lambda_{i_1}|^2 + \cdots + |\lambda_{i_n}|^2,$$

whence $\lambda_{i_1} = \cdots = \lambda_{i_n} = 0$.

If E has finite dimension p, the cardinal number of I is therefore at most equal to p. However, in general there exist infinite orthonormal families. For example in l^2, let $e_n = (0, 0, \ldots, 0, 1, 0, 0, \ldots)$, where the 1 appears in the n'th place. Then the sequence $(e_1, e_2, \ldots)$ is orthonormal.

8.8.7. Let E be a separated pre-Hilbert space. One calls *orthonormal basis* of E a total orthonormal family in E, that is (8.7.4), an orthonormal family $(e_i)_{i\in \mathrm{I}}$ such that the linear combinations of the e_i are dense in E.

Suppose E is finite-dimensional. By 8.7.1, an orthonormal basis of E is an orthonormal family that is a basis in the algebraic sense. For example, in $\mathbf{C}^n$ equipped with the canonical scalar product, the canonical basis is an orthonormal basis.

In general, an orthonormal basis is not a basis in the algebraic sense (but there is almost never any risk of confusion). For example in l^2, the sequence (e_n) of 8.8.6 is an orthonormal basis by 8.7.1.

► **8.8.8. Theorem.** *Consider* $\mathrm{E} = \mathscr{C}([0, 1], \mathbf{C})$ *as a separated pre-Hilbert space with the scalar product* $(f, g) \mapsto \int_0^1 \overline{f(t)}g(t)\,dt$ (8.4.5). *For every* $n \in \mathbf{Z}$, *let* e_n *be the function* $t \mapsto e^{2\pi int}$ *on* $[0, 1]$, *which is an element of* E. *Then the family* $(e_n)_{n\in\mathbf{Z}}$ *is an orthonormal basis of* E.

(a) We have $(e_m\,|\,e_n) = \int_0^1 e^{-2\pi imt}e^{2\pi int}\,dt = \int_0^1 e^{2\pi i(n-m)t}\,dt$. If $n = m$, the value of the integral is 1. If $n \neq m$, then the function $t \mapsto e^{2\pi i(n-m)t}$ admits the primitive $e^{2\pi i(n-m)t}/2\pi i(n - m)$, which takes on the same value at $t = 0$ and $t = 1$, therefore the integral is 0. Thus, the family $(e_n)_{n\in\mathbf{Z}}$ is orthonormal.

(b) Let E′ be the set of $f \in \mathrm{E}$ such that $f(0) = f(1)$. If $f \in \mathrm{E}'$ then f may be extended in a unique way to a function g of period 1 on $\mathbf{R}$, and g is continuous. By 7.5.6, there exists a sequence (f_p) of trigonometric polynomials that tends to g uniformly on $\mathbf{R}$. Then $\int_0^1 |f(t) - f_p(t)|^2\,dt \to 0$, thus (f_p) tends to f in the pre-Hilbert space E. Now, f_p is a finite linear combination of the e_n. Thus, if we denote by A the linear subspace of E generated by the e_n, then $\overline{\mathrm{A}} \supset \mathrm{E}'$.

(c) Let $h \in E$. For $n = 1, 2, 3, \ldots$ let h_n be the function that coincides with h on $[1/n, 1]$, such that $h_n(0) = h_n(1)$, and which is linear on $[0, 1/n]$. The $|h_n|$ are bounded above by a fixed constant M (for example, $M = \sup_{t \in [0,1]} |h(t)|$). Then

$$d(h, h_n)^2 = \int_0^1 |h(t) - h_n(t)|^2 \, dt$$

$$= \int_0^{1/n} |h(t) - h_n(t)|^2 \, dt \leq \frac{1}{n} (2M)^2,$$

therefore h_n tends to h in E. Now, $h_n \in E'$. Therefore $\overline{E'} = E$. Then $\overline{A} \supset \overline{E'} = E$, whence $\overline{A} = E$, and the family (e_n) is an orthonormal basis.

*▶ **8.8.9. Theorem.** *Every Hilbert space has an orthonormal basis.*

Let E be a Hilbert space. Consider the orthonormal subsets of E. They form a set $\mathscr{P}$ ordered by inclusion. Let (P_λ) be a totally ordered family of elements of $\mathscr{P}$. Each P_λ is an orthonormal subset of E. Moreover, for any λ and μ, either $P_\lambda \supset P_\mu$ or $P_\mu \supset P_\lambda$; it is then clear that the union of the P_λ is an orthonormal set, containing all the P_λ. By Zorn's theorem, there exists a maximal orthonormal subset P of E. If P is not total in E, then there exists a nonzero x in E orthogonal to P (8.8.4); replacing x by $x/\|x\|$, one can suppose $\|x\| = 1$. Then $P \cup \{x\}$ is an orthonormal subset of E, which contradicts the maximality of P. Thus P is total in E, hence is an orthonormal basis.

CHAPTER IX

Infinite Sums

The student already knows the definition of a convergent series $x_1 + x_2 + \cdots$ of real numbers, and the definition of the sum of such a series. In this chapter, we generalize in two different directions:

(1) Instead of the x_i being real numbers, we take them to be vectors in a normed space. This is a fairly superficial generalization (though useful in certain contexts—see §5).

(2) Instead of the x_i being indexed by the integers $1, 2, \ldots$, we assume that the set of indices is arbitrary. A lot of very concrete questions lead in fact to the case where the set of indices is $\mathbf{N}^2$ ('double series'), or $\mathbf{N}^p$ ('p-fold series'), or $\mathbf{Z}$, etc., and the best thing is to study at one stroke the general situation. We shall then see why we spoke of 'limit along a filtering set' in 7.2.3.

Although the case of series is a good point of reference, it is well to be prudent: for example, compare Theorem 9.4.6 with the well-known fact that a convergent series is not always absolutely convergent.

9.1. Summable Families

9.1.1. Definition. Let E be a normed space, $(x_i)_{i \in \mathrm{I}}$ a family of elements of E. Let Λ be the set of finite subsets of I; it is an increasingly filtering ordered set, thus one can speak of limit along Λ (7.2.3). For every $\mathrm{J} \in \Lambda$, let

$$s_{\mathrm{J}} = \sum_{i \in \mathrm{J}} x_i \in \mathrm{E}.$$

Let $s \in \mathrm{E}$.

The family $(x_i)_{i \in I}$ is said to be *summable*, with *sum* s, if the family $(s_J)_{J \in \Lambda}$ tends to s along Λ. We then write $s = \sum_{i \in I} x_i$.

In other words, $(x_i)_{i \in I}$ is summable with sum s if, for every $\varepsilon > 0$, there exists a finite subset J_0 of I such that, for every finite subset J of I containing J_0, one has $\|\sum_{i \in J} x_i - s\| \leq \varepsilon$.

9.1.2. Theorem. *Let $(x_i)_{i \in I}$, $(y_i)_{i \in I}$ be summable families in E, with sums s, t. Then the family $(x_i + y_i)_{i \in I}$ is summable with sum $s + t$.*

For $J \in \Lambda$, let $s_J = \sum_{i \in J} x_i$, $t_J = \sum_{i \in J} y_i$. Then, along Λ, s_J tends to s and t_J tends to t, therefore $(s_J, t_J) \in E \times E$ tends to $(s, t) \in E \times E$, therefore $s_J + t_J$ tends to $s + t$. Now, $s_J + t_J = \sum_{i \in J} (x_i + y_i)$.

9.1.3. Similarly, if λ is a scalar, then the family $(\lambda x_i)_{i \in I}$ is summable with sum λs. We thus have the formulas

$$\sum_{i \in I} (x_i + y_i) = \sum_{i \in I} x_i + \sum_{i \in I} y_i, \qquad \sum_{i \in I} \lambda x_i = \lambda \sum_{i \in I} x_i.$$

9.1.4. Theorem. *Let $(x_i)_{i \in I}$ be a summable infinite family of elements of* E. *Then x_i tends to 0 along the filter of complements of finite subsets of* I.

Let $s = \sum_{i \in I} x_i$. Let $\varepsilon > 0$. There exists $J \in \Lambda$ such that

$$J' \in \Lambda,\ J' \supset J \Rightarrow \|s_{J'} - s\| \leq \frac{\varepsilon}{2}.$$

Then

$$i \in I - J \Rightarrow \|s_{J \cup \{i\}} - s\| \leq \frac{\varepsilon}{2} \quad \text{and} \quad \|s_J - s\| \leq \frac{\varepsilon}{2}$$

$$\Rightarrow \|s_{J \cup \{i\}} - s_J\| \leq \varepsilon \Rightarrow \|x_i\| \leq \varepsilon.$$

9.1.5. Example. If $(x_1, x_2, \ldots)$ is a summable sequence of elements of E, then the sequence (x_n) tends to 0 as $n \to \infty$. It is well known, from the example $E = \mathbf{R}$, that the converse is not true.

► **9.1.6. Theorem** (Cauchy's Criterion). *Let E be a Banach space, $(x_i)_{i \in I}$ a family of elements of* E. *The following conditions are equivalent:*

(i) *the family $(x_i)_{i \in I}$ is summable;*
(ii) *for every $\varepsilon > 0$, there exists a finite subset J_0 of I such that, for every finite subset K of I disjoint from J_0, one has $\|s_K\| \leq \varepsilon$.*

Suppose $(x_i)_{i\in I}$ is summable with sum s. Let $\varepsilon > 0$. There exists $J_0 \in \Lambda$ such that, if $J \in \Lambda$ and $J \supset J_0$, then $\|s_J - s\| \leq \varepsilon/2$. Let $K \in \Lambda$ with $K \cap J_0 = \varnothing$. Then

$$\|s_{J_0} - s\| \leq \frac{\varepsilon}{2} \quad \text{and} \quad \|s_{J_0 \cup K} - s\| \leq \frac{\varepsilon}{2},$$

therefore $\|s_K\| = \|s_{J_0\cup K} - s_{J_0}\| \leq \varepsilon$.

Suppose that the condition (ii) is satisfied. Let $\varepsilon > 0$. There exists $J_0 \in \Lambda$ such that, for $K \in \Lambda$ satisfying $K \cap J_0 = \varnothing$, one has $\|s_K\| \leq \varepsilon/2$. If $J, J' \in \Lambda$ and $J, J' \supset J_0$, then

$$\|s_J - s_{J_0}\| \leq \frac{\varepsilon}{2}, \quad \|s_{J'} - s_{J_0}\| \leq \frac{\varepsilon}{2},$$

therefore $\|s_J - s_{J'}\| \leq \varepsilon$. Thus, the set of s_J, for $J \in \Lambda$ and $J \supset J_0$, has diameter $\leq \varepsilon$. Therefore $(s_J)_{J\in\Lambda}$ has a limit along Λ (5.5.11).

9.1.7. Definition. Let E be a normed space, $(x_i)_{i\in I}$ a family of elements of E. The family $(x_i)_{i\in I}$ is said to be *absolutely summable* if the family $(\|x_i\|)_{i\in I}$ is summable in **R**.

► **9.1.8. Theorem.** *Let* E *be a Banach space,* $(x_i)_{i\in I}$ *an absolutely summable family of elements of* E. *Then* $(x_i)_{i\in I}$ *is summable.*

Let $\varepsilon > 0$. There exists $J_0 \in \Lambda$ such that

$$K \in \Lambda,\ K \cap J_0 = \varnothing \ \Rightarrow\ \sum_{i\in K} \|x_i\| \leq \varepsilon \ \Rightarrow\ \left\|\sum_{i\in K} x_i\right\| \leq \varepsilon,$$

therefore $(x_i)_{i\in I}$ is summable by 9.1.6.

9.2. Associativity, Commutativity

9.2.1. Theorem. *Let* E *be a Banach space,* $(x_i)_{i\in I}$ *a summable family of elements of* E. *Let* $J \subset I$. *Then* $(x_i)_{i\in J}$ *is summable.*

Let $\varepsilon > 0$. There exists a finite subset J_0 of I such that, if K is a finite subset of I disjoint from J_0, then $\|s_K\| \leq \varepsilon$. Then $J \cap J_0$ is a finite subset of J, and if K′ is a finite subset of J disjoint from $J \cap J_0$ then $\|s_{K'}\| \leq \varepsilon$. Cauchy's criterion applied to $(x_i)_{i\in J}$ proves that this family is summable.

► **9.2.2. Theorem** (Associativity). *Let* E *be a Banach space,* $(x_i)_{i\in I}$ *a summable family in* E, $(I_l)_{l\in L}$ *a partition of* I. *For every* $l \in L$, *set* $y_l = \sum_{i\in I_l} x_i$, *which is meaningful by* 9.2.1. *Then the family* $(y_l)_{l\in L}$ *is summable and* $\sum_{i\in I} x_i = \sum_{l\in L} y_l$.

Set $s = \sum_{i \in I} x_i$. Let $\varepsilon > 0$. There exists a finite subset J_0 of I such that, if J is a finite subset of I containing J_0, then $\|s_J - s\| \leq \varepsilon/2$.

Let $M_0 \subset L$ be the set of $l \in L$ such that I_l intersects J_0. This is a finite subset of L. Let M be a finite subset of L containing M_0. We are going to show that

$$\left\| \sum_{l \in M} y_l - s \right\| \leq \varepsilon, \tag{1}$$

which will establish the theorem.

Let n be the number of elements of M. For each $l \in L$, there exists a finite subset I'_l of I_l such that $\|y_l - s_{I'_l}\| \leq \varepsilon/2n$, and we can require that I'_l contain $J_0 \cap I_l$ by enlarging it if necessary. The union of the I'_l, as l runs over M, is a finite subset J of I, and $J \supset J_0$. We have $\|s_J - s\| \leq \varepsilon/2$, that is,

$$\left\| \sum_{l \in M} s_{I'_l} - s \right\| \leq \varepsilon/2. \tag{2}$$

Now, $\|y_l - s_{I'_l}\| \leq \varepsilon/2n$ for every $l \in M$, therefore

$$\left\| \sum_{l \in M} y_l - \sum_{l \in M} s_{I'_l} \right\| \leq \varepsilon/2. \tag{3}$$

The inequality (1) follows from (2) and (3).

9.2.3. Example. Let $(x_{mn})_{m,n=1,2,\ldots}$ be a double sequence of real numbers. If it is summable, then

$$\begin{aligned} \sum_{m,n \geq 1} x_{mn} &= \sum_{m \geq 1} \left(\sum_{n \geq 1} x_{mn} \right) \\ &= \sum_{n \geq 1} \left(\sum_{m \geq 1} x_{mn} \right). \end{aligned}$$

9.2.4. Let E be a Banach space, $(x_i)_{i \in I}$ a family of elements of E, $(I_l)_{l \in L}$ a partition of I. Suppose that each subfamily $(x_i)_{i \in I_l}$ is summable with sum y_l, and that the family $(y_l)_{l \in L}$ is summable. This does *not* imply that the family $(x_i)_{i \in I}$ is summable. For example, take $E = \mathbf{R}$ and consider the sequence of real numbers $(1, -1, 2, -2, \ldots, n, -n, \ldots)$. Each subfamily $(n, -n)$ is summable with sum 0, and the family $(0, 0, 0, \ldots)$ is summable, yet the original sequence is not summable (for example by 9.1.5).

9.2.5. However, one has the following result:

Theorem. *Let* E *be a normed vector space,* $(x_i)_{i \in I}$ *a family of elements of* E*,* $(I_l)_{l \in L}$ *a partition of* I *with* L *finite. Assume that each subfamily* $(x_i)_{i \in I_l}$ *is summable with sum* y_l*. Then* $(x_i)_{i \in I}$ *is summable with sum* $\sum_{l \in L} y_l$.

Let $\varepsilon > 0$. Let n be the number of elements of L. For every $l \in \mathrm{L}$, there exists a finite subset J_l of I_l such that, if J′ is a finite subset of I_l containing J_l, then $\|s_{\mathrm{J}'} - y_l\| \leq \varepsilon/n$. Let $\mathrm{J} = \bigcup_{l \in \mathrm{L}} \mathrm{J}_l$, which is a finite subset of I. If J′ is a finite subset of I containing J, then J′ is the disjoint union of the $\mathrm{J}' \cap \mathrm{I}_l$, and $\mathrm{J}' \cap \mathrm{I}_l$ is a finite subset of I_l containing J_l. Therefore $\|s_{\mathrm{J}' \cap \mathrm{I}_l} - y_l\| \leq \varepsilon/n$, and this for every $l \in \mathrm{L}$. Since $s_{\mathrm{J}'} = \sum_{l \in \mathrm{L}} s_{\mathrm{J}' \cap \mathrm{I}_l}$, we have $\|s_{\mathrm{J}'} - \sum_{l \in \mathrm{L}} y_l\| \leq \varepsilon$, whence the theorem.

9.2.6. Theorem (Commutativity). *Let* E *be a normed space,* $(x_i)_{i \in \mathrm{I}}$ *a summable family of elements of* E *with sum s. Let* σ *be a bijection of* I *onto* I. *Then the family* $(x_{\sigma(i)})_{i \in \mathrm{I}}$ *is summable with sum s.*

Let $\varepsilon > 0$. Let J_0 be as in 9.1.1. Then $\sigma^{-1}(\mathrm{J}_0)$ is a finite subset of I. Let J be a finite subset of I containing $\sigma^{-1}(\mathrm{J}_0)$. Then $\sigma(\mathrm{J}) \supset \mathrm{J}_0$, therefore $\|\sum_{i \in \sigma(\mathrm{J})} x_i - s\| \leq \varepsilon$, that is,

$$\left\| \sum_{i \in \mathrm{J}} x_{\sigma(i)} - s \right\| \leq \varepsilon,$$

whence the theorem.

9.3. Series

9.3.1. Definition. Let E be a normed space, and $(x_1, x_2, \ldots)$ a *sequence* of elements of E. We say that the series with general term x_n is *convergent with sum s* (where $s \in \mathrm{E}$) if $s_n = \sum_{i=1}^{n} x_i$ tends to s as n tends to infinity. We then write $s = \sum_{n=1}^{\infty} x_n$.

In other words, the series is convergent with sum s if, for every $\varepsilon > 0$, there exists an N such that $n \geq \mathrm{N} \Rightarrow \|s_n - s\| \leq \varepsilon$.

The series with general term x_n is said to be *absolutely convergent* if the series with general term $\|x_n\|$ is convergent.

9.3.2. By means of proofs analogous to those of 9.1, one establishes the following results:

(a) If the series with general terms x_n and y_n are convergent, then the series with general terms $x_n + y_n$ and λx_n (λ a scalar) are convergent, and

$$\sum_{n=1}^{\infty} (x_n + y_n) = \sum_{n=1}^{\infty} x_n + \sum_{n=1}^{\infty} y_n,$$

$$\sum_{n=1}^{\infty} \lambda x_n = \lambda \sum_{n=1}^{\infty} x_n.$$

(b) If the series with general term x_n is convergent, then x_n tends to 0 as n tends to infinity. (The converse is not true.)

(c) (Cauchy's Criterion.) Let E be a Banach space, $(x_1, x_2, \ldots)$ a sequence of elements of E. The following conditions are equivalent:

(i) the series with general term x_n is convergent;
(ii) for every $\varepsilon > 0$, there exists an N such that

$$n \geq m \geq \mathrm{N} \Rightarrow \left\| \sum_{i=m}^{n} x_i \right\| \leq \varepsilon.$$

(d) Let E be a Banach space, $(x_1, x_2, \ldots)$ a sequence of elements of E. If the series with general term x_n is absolutely convergent, then the series is convergent. (The converse is not true.)

9.3.3. Theorem. *Let E be a normed space, $(x_1, x_2, \ldots)$ a sequence of elements of E. If the sequence is summable with sum s, then the series with general term x_n is convergent with sum s.*

Let $\varepsilon > 0$. There exists a finite subset J_0 of $\{1, 2, \ldots\}$ such that, if J is a finite subset of $\{1, 2, \ldots\}$ containing J_0, then $\|s - \sum_{i \in \mathrm{J}} x_i\| \leq \varepsilon$. Let N be the largest of the integers in J_0. If $n \geq \mathrm{N}$ then $\mathrm{J}_0 \subset \{1, 2, \ldots, n\}$, therefore $\|s - \sum_{i=1}^{n} x_i\| \leq \varepsilon$. This proves that $\sum_{i=1}^{n} x_i \to s$ as $n \to \infty$.

9.3.4. The converse of 9.3.3 is false, as will follow from the example in 9.3.5.

9.3.5. Let E be a normed space, $(x_1, x_2, \ldots)$ a sequence of elements of E, σ a permutation of $\{1, 2, 3, \ldots\}$. If the sequence $(x_1, x_2, \ldots)$ is summable, then $\sum_{n=1}^{\infty} x_n = \sum_{n=1}^{\infty} x_{\sigma(n)}$ by 9.3.3 and 9.2.6. However, consider the series

$$1 - \tfrac{1}{2} + \tfrac{1}{3} - \tfrac{1}{4} + \tfrac{1}{5} - \cdots,$$

which one knows to be convergent with sum log 2. By a suitable permutation of the order of the terms, one obtains the series

$$\underbrace{1 - \tfrac{1}{2}} - \tfrac{1}{4} + \underbrace{\tfrac{1}{3} - \tfrac{1}{6}} - \tfrac{1}{8} + \underbrace{\tfrac{1}{5} - \tfrac{1}{10}} - \tfrac{1}{12} + \cdots.$$

This series is easily seen to be convergent, with the same sum as

$$\tfrac{1}{2} - \tfrac{1}{4} + \tfrac{1}{6} - \tfrac{1}{8} + \tfrac{1}{10} - \tfrac{1}{12} + \cdots;$$

this sum is therefore $\frac{1}{2}$ log 2. *The sum has changed after rearrangement of the terms.* In particular, the sequence $(1, -\frac{1}{2}, \frac{1}{3}, -\frac{1}{4}, \ldots)$ is not summable (cf. also 9.4.6).

9.4. Summable Families of Real or Complex Numbers

▶ **9.4.1. Theorem.** *Let $(x_i)_{i\in I}$ be a family of real numbers ≥ 0. Consider the finite partial sums $s_J = \sum_{i\in J} x_i$, where $J \in \Lambda$ (the set of all finite subsets of I). Let $s = \sup_{J\in\Lambda} s_J \in [0, +\infty]$.*

(i) *If $s < +\infty$, then the family $(x_i)_{i\in I}$ is summable with sum s.*
(ii) *If $s = +\infty$, then s_J tends to $+\infty$ along Λ.*

The mapping $J \mapsto s_J$ of Λ into **R** is increasing, since the x_i are ≥ 0. The theorem therefore follows from 7.2.4.

9.4.2. Thus, for a family $(x_i)_{i\in I}$ of numbers ≥ 0, the symbol $\sum_{i\in I} x_i$ *is always meaningful*: it is a finite number if the family $(x_i)_{i\in I}$ is summable, $+\infty$ otherwise.

9.4.3. Theorem. *Let $(x_i)_{i\in I}$, $(y_i)_{i\in I}$ be two families of numbers ≥ 0; assume that $x_i \leq y_i$ for all $i \in I$. Then $\sum_{i\in I} x_i \leq \sum_{i\in I} y_i$. In particular, if the family $(y_i)_{i\in I}$ is summable, then the family $(x_i)_{i\in I}$ is summable.*

Let Λ be the set of finite subsets of I. If $J \in \Lambda$, then $\sum_{i\in J} x_i \leq \sum_{i\in J} y_i$. Passing to the limit along Λ, we obtain $\sum_{i\in I} x_i \leq \sum_{i\in I} y_i$.

9.4.4. Theorem (Associativity). *Let $(x_i)_{i\in I}$ be a family of numbers ≥ 0, $(I_l)_{l\in L}$ a partition of I. Then*

$$\sum_{i\in I} x_i = \sum_{l\in L}\left(\sum_{i\in I_l} x_i\right).$$

If the family $(x_i)_{i\in I}$ is summable, this follows from 9.2.2. Otherwise, $\sum_{i\in I} x_i = +\infty$. For every finite subset J of I, we have

$$\sum_{i\in J} x_i = \sum_{l\in L}\left(\sum_{i\in I_l\cap J} x_i\right)$$

$$\leq \sum_{l\in L}\left(\sum_{i\in I_l} x_i\right) \quad \text{by 9.4.3.}$$

This being true for all J, we conclude that

$$\sum_{l\in L}\left(\sum_{i\in I_l} x_i\right) = +\infty,$$

thus the equality of the theorem is again true.

9.4.5. Theorem. *Let* $(x_i)_{i \in \mathrm{I}}$, $(y_j)_{j \in \mathrm{J}}$ *be families of numbers* ≥ 0. *Then*

$$\sum_{(i, j) \in \mathrm{I} \times \mathrm{J}} x_i y_j = \left(\sum_{i \in \mathrm{I}} x_i\right)\left(\sum_{j \in \mathrm{J}} y_j\right).$$

(We make the convention that $0 \cdot +\infty = 0$.)

By 9.4.4, we have

$$\sum_{(i, j) \in \mathrm{I} \times \mathrm{J}} x_i y_j = \sum_{i \in \mathrm{I}} \left(\sum_{j \in \mathrm{J}} x_i y_j\right).$$

Now, $\sum_{j \in \mathrm{J}} x_i y_j = x_i \sum_{j \in \mathrm{J}} y_j$. Setting $t = \sum_{j \in \mathrm{J}} y_j$, we thus have

$$\sum_{(i, j) \in \mathrm{I} \times \mathrm{J}} x_i y_j = \sum_{i \in \mathrm{I}} x_i t.$$

It $t < +\infty$, this is equal to $t(\sum_{i \in \mathrm{I}} x_i)$, whence the theorem. If $t = +\infty$, we distinguish two cases. If all of the x_i are zero, then $tx_i = 0$ for all i, therefore $\sum_{i \in \mathrm{I}} tx_i = 0 = t \sum_{i \in \mathrm{I}} x_i$. If one of the x_i is > 0, then $tx_i = +\infty$ for such an i, therefore $\sum_{i \in \mathrm{I}} tx_i = +\infty$; on the other hand, $\sum_{i \in \mathrm{I}} x_i > 0$, therefore $t \sum_{i \in \mathrm{I}} x_i = +\infty$.

► **9.4.6. Theorem.** *Let* $(x_i)_{i \in \mathrm{I}}$ *be a family of real or complex numbers. The following conditions are equivalent*:

(i) *the family* $(x_i)_{i \in \mathrm{I}}$ *is summable*;
(ii) *the family* $(x_i)_{i \in \mathrm{I}}$ *is absolutely summable.*

(ii) $\Rightarrow$ (i). This follows from 9.1.8.

(i) $\Rightarrow$ (ii). Suppose that the family $(x_i)_{i \in \mathrm{I}}$ is summable. If the x_i are all real, let I_1 (resp. I_2) be the set of $i \in \mathrm{I}$ such that $x_i \geq 0$ (resp. $x_i < 0$). The families $(x_i)_{i \in \mathrm{I}_1}$ and $(x_i)_{i \in \mathrm{I}_2}$ are summable (9.2.1). Therefore the families $(|x_i|)_{i \in \mathrm{I}_1}$ and $(|x_i|)_{i \in \mathrm{I}_2}$ are summable (9.1.3). Therefore the family $(|x_i|)_{i \in \mathrm{I}}$ is summable (9.2.5). If the x_i are complex, then the family $(\bar{x}_i)$ is clearly summable. Since $\operatorname{Re} x_i = \frac{1}{2}(x_i + \bar{x}_i)$, the family $(\operatorname{Re} x_i)$ is summable (9.1.2). Similarly, the family $(\operatorname{Im} x_i)$ is summable. Then the families $(|\operatorname{Re} x_i|)$, $(|\operatorname{Im} x_i|)$ are summable by the first part of the proof, therefore so is the family $(|\operatorname{Re} x_i| + |\operatorname{Im} x_i|)$. Finally, since $|x_i| \leq |\operatorname{Re} x_i| + |\operatorname{Im} x_i|$, the family $(|x_i|)$ is summable (9.4.3).

9.4.7. Series with terms ≥ 0. If $x_1, x_2, \ldots$ are numbers ≥ 0, then the sequence $(\sum_{i=1}^{n} x_i)$ is increasing, therefore has a limit in $\overline{\mathbf{R}}$ equal to its supremum s, which is denoted $\sum_{i=1}^{\infty} x_i$. *This number is equal to* $\sum_{i \in \{1, 2, \ldots\}} x_i$. For, if the sequence (x_i) is summable, this follows from 9.3.3. Otherwise, there exist arbitrarily large finite subsums $\sum_{i \in \mathrm{J}} x_i$ (9.4.1), therefore arbitrarily large sums $\sum_{i=1}^{n} x_i$, therefore

$$\sum_{i=1}^{\infty} x_i = +\infty = \sum_{i \in \{1, 2, \ldots\}} x_i.$$

9.4.8. To sum up, if $(x_i)_{i \in I}$ is a family of elements of a normed space E, one has the following diagrams:

I arbitrary, E complete:

$$\text{absolute summability} \Rightarrow \text{summability}.$$

I arbitrary, $E = \mathbf{R}$ or $\mathbf{C}$:

$$\text{absolute summability} \Leftrightarrow \text{summability}.$$

$I = \{1, 2, \ldots\}$, E complete:

$$\begin{array}{ccc} \text{absolute summability} & \Rightarrow & \text{summability} \\ \Updownarrow & & \Downarrow \\ \text{absolute convergence} & \Rightarrow & \text{convergence.} \end{array}$$

$I = \{1, 2, \ldots\}$, $E = \mathbf{R}$ or $\mathbf{C}$:

$$\begin{array}{ccc} \text{absolute summability} & \Leftrightarrow & \text{summability} \\ \Updownarrow & & \Downarrow \\ \text{absolute convergence} & \Rightarrow & \text{convergence.} \end{array}$$

9.4.9. Let $\alpha \in \mathbf{R}$. One knows that the number

$$1 + \frac{1}{2^\alpha} + \frac{1}{3^\alpha} + \frac{1}{4^\alpha} + \cdots = \sum_{x \in \mathbf{N} - \{0\}} \frac{1}{x^\alpha}$$

is finite if $\alpha > 1$ and infinite if $\alpha \leq 1$. This amounts to saying that

$$\sum_{x \in \mathbf{Z} - \{0\}} \frac{1}{|x|^\alpha} < +\infty \Leftrightarrow \alpha > 1.$$

Here is an important generalization:

Theorem. *Let $x \mapsto \|x\|$ be a norm on $\mathbf{R}^p$ and let $\alpha \in \mathbf{R}$. Then*

$$\sum_{x \in \mathbf{Z}^p - \{0\}} \frac{1}{\|x\|^\alpha} < +\infty \Leftrightarrow \alpha > p.$$

Since all norms on $\mathbf{R}^p$ are equivalent, it suffices (in view of 9.4.3) to carry out the proof for the norm

$$(x_1, x_2, \ldots, x_p) \mapsto \sup(|x_1|, |x_2|, \ldots, |x_p|).$$

On $\mathbf{Z}^p - \{0\}$, this norm takes on the values $1, 2, 3, \ldots$. For $n = 1, 2, 3, \ldots$, let A_n be the set of $x \in \mathbf{Z}^p$ such that $\|x\| = n$. Then

$$A_n = B_{n1} \cup B_{n2} \cup \cdots \cup B_{np},$$

where B_{ni} is the set of $(x_1, x_2, \ldots, x_p) \in \mathbf{Z}^p$ such that $x_i = \pm n$ and $-n \leq x_j \leq n$ for $j \neq i$. It is clear that $\mathrm{Card}(B_{ni}) = 2(2n+1)^{p-1}$, therefore

$$n^{p-1} \leq \mathrm{Card}\, A_n \leq 2p(2n+1)^{p-1} \leq 2p(3n)^{p-1} = 2 \cdot 3^{p-1} \cdot pn^{p-1}.$$

Consequently,

$$\frac{1}{n^{\alpha-p+1}} = \frac{n^{p-1}}{n^{\alpha}} \leq \sum_{x\in A_n} \frac{1}{\|x\|^{\alpha}} \leq \frac{2\cdot 3^{p-1}\cdot pn^{p-1}}{n^{\alpha}} = 2\cdot 3^{p-1}\cdot p\cdot\frac{1}{n^{\alpha-p+1}}.$$

By 9.4.3 and 9.4.4, we deduce from this that

$$\sum_{n\in\mathbf{N}-\{0\}} \frac{1}{n^{\alpha-p+1}} \leq \sum_{x\in\mathbf{Z}^p-\{0\}} \frac{1}{\|x\|^{\alpha}} \leq 2\cdot 3^{p-1}\cdot p \sum_{n\in\mathbf{N}-\{0\}} \frac{1}{n^{\alpha-p+1}},$$

from which

$$\sum_{x\in\mathbf{Z}^p-\{0\}} \frac{1}{\|x\|^{\alpha}} < +\infty \Leftrightarrow \sum_{n\in\mathbf{N}-\{0\}} \frac{1}{n^{\alpha-p+1}} < +\infty$$

$$\Leftrightarrow \alpha - p + 1 > 1 \Leftrightarrow \alpha > p.$$

9.4.10. The Hilbert Space $l^2(\mathrm{I})$. Let I be a set. We denote by $l^2_{\mathbf{C}}(\mathrm{I})$, or simply $l^2(\mathrm{I})$, the set of families $x = (x_i)_{i\in\mathrm{I}}$ of complex numbers such that $\sum_{i\in\mathrm{I}} |x_i|^2 < +\infty$. Let us show that this is a linear subspace of $\mathscr{F}(\mathrm{I}, \mathbf{C})$. It is clear that if $x \in l^2(\mathrm{I})$ and $\lambda \in \mathbf{C}$, then $\lambda x \in l^2(\mathrm{I})$. Let $x = (x_i)_{i\in\mathrm{I}} \in l^2(\mathrm{I})$ and $y = (y_i)_{i\in\mathrm{I}} \in l^2(\mathrm{I})$. Then

$$\sum_{i\in\mathrm{I}} |x_i + y_i|^2 \leq \sum_{i\in\mathrm{I}} (2|x_i|^2 + 2|y_i|^2) \quad \text{by 9.4.3}$$

$$= 2\sum_{i\in\mathrm{I}} |x_i|^2 + 2\sum_{i\in\mathrm{I}} |y_i|^2 \quad \text{by 9.1.2, 9.1.3}$$

$$< +\infty,$$

thus $x + y \in l^2(\mathrm{I})$. Next,

$$2\sum_{i\in\mathrm{I}} |x_i||y_i| \leq \sum_{i\in\mathrm{I}} (|x_i|^2 + |y_i|^2) < +\infty,$$

so that the family $(\bar{x}_i y_i)_{i\in\mathrm{I}}$ is absolutely summable, therefore summable (9.1.8). Set $(x|y) = \sum_{i\in\mathrm{I}} \bar{x}_i y_i$. One verifies easily that $(x, y) \mapsto (x|y)$ is a scalar product on $l^2(\mathrm{I})$. Imitating 8.6.4 step by step, one sees that $l^2(\mathrm{I})$ *is a Hilbert space.* (In the proof of 8.6.4, consideration of the sum $\sum_{i=1}^{\Lambda}$ must be replaced by that of a sum $\sum_{i\in\mathrm{J}}$, where J is a finite subset of I.)

If $\mathrm{I} = \{1, 2, 3, \ldots\}$, then $l^2(\mathrm{I}) = l^2$. If $\mathrm{I} = \{1, 2, \ldots, n\}$, then $l^2(\mathrm{I}) = \mathbf{C}^n$ equipped with the canonical scalar product.

Let $i_0 \in \mathrm{I}$. Consider the family $e_{i_0} = (x_i)_{i\in\mathrm{I}}$ such that $x_i = 0$ for $i \neq i_0$, $x_i = 1$ for $i = i_0$. Then $e_{i_0} \in l^2(\mathrm{I})$. The family $(e_i)_{i\in\mathrm{I}}$ in $l^2(\mathrm{I})$ is orthonormal. As in 8.7.1, one verifies that the linear combinations of the e_i are dense in $l^2(\mathrm{I})$, thus $(e_i)_{i\in\mathrm{I}}$ is an orthonormal basis of $l^2(\mathrm{I})$, called the *canonical orthonormal basis.*

9.5. Certain Summable Families in Hilbert Spaces

► **9.5.1. Theorem.** *Let* E *be a separated pre-Hilbert space,* $(x_i)_{i \in \mathrm{I}}$ *a family of pairwise orthogonal elements of* E.

(i) *If the family* $(x_i)_{i \in \mathrm{I}}$ *is summable with sum* s, *then*

$$\sum_{i \in \mathrm{I}} \|x_i\|^2 < +\infty \quad \text{and} \quad \|s\|^2 = \sum_{i \in \mathrm{I}} \|x_i\|^2.$$

(ii) *If* $\sum_{i \in \mathrm{I}} \|x_i\|^2 < +\infty$ *and* E *is a Hilbert space, then the family* $(x_i)_{i \in \mathrm{I}}$ *is summable.*

Let Λ be the set of finite subsets of I. For every $\mathrm{J} \in \Lambda$, let $s_{\mathrm{J}} = \sum_{i \in \mathrm{J}} x_i$.

If $(x_i)_{i \in \mathrm{I}}$ is summable with sum s, then s_{J} tends to s along Λ, therefore $\|s_{\mathrm{J}}\|^2$ tends to $\|s\|^2$ along Λ. Now, $\|s_{\mathrm{J}}\|^2 = \sum_{i \in \mathrm{J}} \|x_i\|^2$ (8.5.7), therefore $\|s\|^2 = \sum_{i \in \mathrm{I}} \|x_i\|^2$.

Suppose $\sum_{i \in \mathrm{I}} \|x_i\|^2 < +\infty$. For every $\varepsilon > 0$, there exists $\mathrm{J} \in \Lambda$ such that, if $\mathrm{K} \in \Lambda$ and $\mathrm{K} \cap \mathrm{J} = \varnothing$, then $\sum_{i \in \mathrm{K}} \|x_i\|^2 \leq \varepsilon^2$ (9.1.6), that is, $\|s_{\mathrm{K}}\| \leq \varepsilon$. If, moreover, E is a Hilbert space, this implies that $(x_i)_{i \in \mathrm{I}}$ is summable (9.1.6).

► **9.5.2. Theorem.** *Let* E *be a separated pre-Hilbert space,* $(e_i)_{i \in \mathrm{I}}$ *an orthonormal basis of* E.

(i) *Let* $x \in \mathrm{E}$, *and set* $\lambda_i = (e_i | x)$. *The family* $(\lambda_i e_i)_{i \in \mathrm{I}}$ *is summable in* E, *and* $x = \sum_{i \in \mathrm{I}} \lambda_i e_i$.

(ii) *If* $y \in \mathrm{E}$ *and* $\mu_i = (e_i | y)$, *then the family* $(\bar{\lambda}_i \mu_i)_{i \in \mathrm{I}}$ *is summable and* $(x | y) = \sum_{i \in \mathrm{I}} \bar{\lambda}_i \mu_i$. *In particular,* $\|x\|^2 = \sum_{i \in \mathrm{I}} |\lambda_i|^2$.

Let $\varepsilon > 0$. There exist a finite subset J_0 of I and a linear combination x' of the e_i for $i \in \mathrm{J}_0$, such that $\|x - x'\| \leq \varepsilon$. Let J be a finite subset of I such that $\mathrm{J} \supset \mathrm{J}_0$, and set

$$z = \sum_{i \in \mathrm{J}} \lambda_i e_i.$$

For $j \in \mathrm{J}$, we have

$$(e_j | x - z) = (e_j | x) - \left(e_j \,\middle|\, \sum_{i \in \mathrm{J}} \lambda_i e_i\right) = \lambda_j - \lambda_j = 0,$$

therefore $x' - z$ (which is a linear combination of the e_j for $j \in \mathrm{J}$) is orthogonal to $x - z$. Consequently

$$\|x - z\|^2 + \|z - x'\|^2 = \|x - x'\|^2 \leq \varepsilon^2,$$

whence $\|x - z\| \leq \varepsilon$. This proves that the family $(\lambda_i e_i)_{i \in \mathrm{I}}$ is summable with sum x.

Since $\sum_{i\in J}\lambda_i e_i$ and $\sum_{i\in J}\mu_i e_i$ tend to x and y along Λ, the number $(\sum_{i\in J}\lambda_i e_i|\sum_{i\in J}\mu_i e_i) = \sum_{i\in J}\bar{\lambda}_i\mu_i$ tends to $(x|y)$ along Λ.

9.5.3. Under the conditions of 9.5.2, we say that the λ_i are the *coordinates of x with respect to the orthonormal basis* $(e_i)_{i\in I}$. If E is finite-dimensional, we recover the usual concept of coordinates with respect to the basis $(e_i)_{i\in I}$, since $x = \sum_{i\in I}\lambda_i e_i$.

► **9.5.4. Corollary.** *Let $f\in\mathscr{C}([0,1],\mathbf{C})$. For every $k\in\mathbf{Z}$, set*

$$\lambda_k = \int_0^1 f(t)e^{-2\pi ikt}\,dt$$

(*'Fourier coefficients of f'*). *Then $\sum_{k=-n}^{n}\lambda_k e^{2\pi ikt}$ tends in mean square to f, as n tends to infinity. If also $g\in\mathscr{C}([0,1],\mathbf{C})$ and $\mu_k = \int_0^1 g(t)e^{-2\pi ikt}\,dt$, then*

$$\int_0^1 f(t)\overline{g(t)}\,dt = \sum_{n\in\mathbf{Z}}\lambda_k\bar{\mu}_k.$$

This follows from 8.8.8 and 9.5.2.

► **9.5.5. Theorem.** *Let* E *be a Hilbert space, $(e_i)_{i\in I}$ an orthonormal basis of* E. *For every $x = (\mu_i)_{i\in I}\in l^2(\mathrm{I})$, let $f(x)$ be the element $\sum_{i\in I}\mu_i e_i$ of* E *(which is defined, by* 9.5.1(ii)). *Then the coordinates of $f(x)$ with respect to $(e_i)_{i\in I}$ are the μ_i, and f is an isomorphism of the Hilbert space $l^2(\mathrm{I})$ onto the Hilbert space* E *that transforms the canonical orthonormal basis of $l^2(\mathrm{I})$ into $(e_i)_{i\in I}$.*

(a) Let Λ be the set of finite subsets of I. Let $j\in\mathrm{I}$. If $\mathrm{J}\in\Lambda$ and $\mathrm{J}\supset\{j\}$, then $(e_j|\sum_{i\in J}\mu_i e_i) = \mu_j$; since $\sum_{i\in J}\mu_i e_i$ tends to $f(x)$ along Λ, we see that $(e_j|f(x))$ is equal to μ_j.

(b) It is clear that f is a linear mapping of $l^2(\mathrm{I})$ into E. For every $y\in\mathrm{E}$, let $g(y)$ be the family of numbers $((e_i|y))_{i\in I}$, which belongs to $l^2(\mathrm{I})$ by 9.5.2. Then g is a mapping of E into $l^2(\mathrm{I})$, and $f(g(y)) = y$ by 9.5.2. On the other hand, $g(f(x)) = x$ for all $x\in l^2(\mathrm{I})$ by (a).

(c) Thus, f and g are linear bijections inverse to one another. By 9.5.2, g preserves the scalar product. Thus f is an isomorphism of the Hilbert space $l^2(\mathrm{I})$ onto the Hilbert space E.

(d) It is clear that f transforms the canonical orthonormal basis of $l^2(\mathrm{I})$ into $(e_i)_{i\in I}$.

*► **9.5.6. Corollary.** *Every Hilbert space is isomorphic to a space $l^2(\mathrm{I})$.*

This follows from 8.8.9 and 9.5.5.

CHAPTER X

Connected Spaces

This chapter, which is very easy, could have come before Chapter III (but there were so many questions urgently requiring study!). The problem is to distinguish, by various methods, those spaces that are 'in one piece' (for example a disc, or the complement of a disc in a plane) and those which are not (for example, the complement of a circle in a plane).

10.1. Connected Spaces

10.1.1. Theorem. *Let* E *be a topological space. The following conditions are equivalent*:

(i) *there exists a subset of* E, *distinct from* E *and* $\varnothing$, *that is both open and closed*;
(ii) *there exist two complementary nonempty subsets of* E *both of which are open*;
(iii) *there exist two complementary nonempty subsets of* E *both of which are closed.*

This is clear, since if A is a subset of E, we have

$$\begin{aligned} \text{A open and closed} &\Leftrightarrow \text{A and E} - \text{A open} \\ &\Leftrightarrow \text{A and E} - \text{A closed.} \end{aligned}$$

10.1.2. Definition. If a topological space E satisfies the conditions of 10.1.1, it is said to be disconnected. In the contrary case, it is said to be *connected.*

10.1.3. Theorem. **R** *is connected.*

Let A be an open and closed subset of **R**. Assuming A and **R** − A nonempty, we are going to arrive at a contradiction. Let $x \in \mathbf{R} - \mathrm{A}$. One of the sets $\mathrm{A} \cap [x, +\infty)$, $\mathrm{A} \cap (-\infty, x]$ is nonempty. Suppose, for example, that $\mathrm{B} = \mathrm{A} \cap [x, +\infty) \neq \varnothing$. Then B is closed. Since $\mathrm{B} = \mathrm{A} \cap (x, +\infty)$, B is also open. Since B is closed, nonempty and bounded below, it has a smallest element b (1.5.9). Since B is open, it contains an interval $(b - \varepsilon, b + \varepsilon)$ with $\varepsilon > 0$. Thus b cannot be the smallest element of B.

10.1.4. However, $\mathbf{R} - \{0\}$ is not connected. For, $(-\infty, 0)$ and $(0, +\infty)$ are complementary nonempty open sets in $\mathbf{R} - \{0\}$.

10.1.5. Definition. Let E be a topological space and $\mathrm{A} \subset \mathrm{E}$. We say that A is a *connected subset* of E if the topological space A is connected.

10.1.6. Theorem. *Let* E *be a topological space,* $(\mathrm{A}_i)_{i \in \mathrm{I}}$ *a family of connected subsets of* E, $\mathrm{A} = \bigcup_{i \in \mathrm{I}} \mathrm{A}_i$. *If the* A_i *intersect pairwise, then* A *is connected.*

Suppose A is not connected. There exist, in the topological space A, subsets U_1, U_2 that are complementary, nonempty and open. For every $i \in \mathrm{I}$, $\mathrm{U}_1 \cap \mathrm{A}_i$ and $\mathrm{U}_2 \cap \mathrm{A}_i$ are open in A_i and complementary in A_i. Since A_i is connected, $\mathrm{U}_1 \cap \mathrm{A}_i = \varnothing$ or $\mathrm{U}_2 \cap \mathrm{A}_i = \varnothing$. Let I_1 (resp. I_2) be the set of $i \in \mathrm{I}$ such that $\mathrm{A}_i \subset \mathrm{U}_1$ (resp. $\mathrm{A}_i \subset \mathrm{U}_2$). Then U_1 (resp. U_2) is the union of the A_i for $i \in \mathrm{I}_1$ (resp. $i \in \mathrm{I}_2$), therefore there exist an A_i and an A_j that are disjoint, contrary to hypothesis.

10.1.7. Theorem. *Let* E *be a topological space,* A *a connected subset of* E, B *a subset of* E *such that* $\mathrm{A} \subset \mathrm{B} \subset \overline{\mathrm{A}}$. *Then* B *is connected. In particular,* $\overline{\mathrm{A}}$ *is connected.*

Suppose B is the union of subsets U_1, U_2 that are disjoint and open in B. We are to prove that one of them is empty. There exist open sets U'_1, U'_2 in E such that $\mathrm{U}_1 = \mathrm{B} \cap \mathrm{U}'_1$, $\mathrm{U}_2 = \mathrm{B} \cap \mathrm{U}'_2$. The sets $\mathrm{A} \cap \mathrm{U}_1$ and $\mathrm{A} \cap \mathrm{U}_2$ are open in A, disjoint, with union A. Since A is connected, we have for example $\mathrm{A} \cap \mathrm{U}_1 = \varnothing$, therefore

$$\mathrm{A} \cap \mathrm{U}'_1 = (\mathrm{A} \cap \mathrm{B}) \cap \mathrm{U}'_1 = \mathrm{A} \cap (\mathrm{B} \cap \mathrm{U}'_1) = \mathrm{A} \cap \mathrm{U}_1 = \varnothing,$$

in other words $\mathrm{A} \subset \mathrm{E} - \mathrm{U}'_1$. Since $\mathrm{E} - \mathrm{U}'_1$ is closed, we infer that $\overline{\mathrm{A}} \subset \mathrm{E} - \mathrm{U}'_1$, whence $\mathrm{B} \cap \mathrm{U}'_1 = \varnothing$, that is, $\mathrm{U}_1 = \varnothing$.

► **10.1.8. Theorem.** *Let* X, Y *be topological spaces,* f *a continuous mapping of* X *into* Y. *If* X *is connected, then* $f(\mathrm{X})$ *is connected.*

If $f(X)$ is not connected, there exist in $f(X)$ sets U_1, U_2 that are open, complementary and nonempty. Then $f^{-1}(U_1)$, $f^{-1}(U_2)$ are open, complementary and nonempty in X, which is absurd.

► **10.1.9. Theorem.** *Let* $A \subset \mathbf{R}$. *The following conditions are equivalent*:

(i) A *is connected*;
(ii) A *is an interval.*

We can suppose that A is nonempty and not reduced to a point.

Let A be an interval. If A is open in **R** then A is homeomorphic to **R** (2.5.5), hence is connected (10.1.3). If A is an arbitrary interval, let I be its interior in **R**. Then I is an open interval in **R** hence is connected, and $I \subset A \subset \bar{I}$, so that A is connected (10.1.7).

Let A be a connected subset of **R** and let us show that A is an interval. By means of the increasing homeomorphism $x \mapsto \tan x$ of $(-\pi/2, \pi/2)$ onto **R**, we are reduced to the case that $A \subset (-\pi/2, \pi/2)$. Then A admits a supremum $b \in \mathbf{R}$ and an infimum $a \in \mathbf{R}$. We have $A \subset [a, b]$. We are going to show that $A \supset (a, b)$; it will then follow that A is one of the four intervals (a, b), $(a, b]$, $[a, b)$, $[a, b]$, and the proof will be complete. Arguing by contradiction, let us suppose that there exists an x_0 such that $a < x_0 < b$ and $x_0 \notin A$. Then A is the union of the sets $A \cap (-\infty, x_0)$ and $A \cap (x_0, +\infty)$, which are open in A. Since A is connected, one of these two sets is empty, say $A \cap (x_0, +\infty)$. Then $x < x_0$ for all $x \in A$, which contradicts the fact that b is the least upper bound of A.

10.1.10. Theorem. *Let* X *be a connected topological space, f a continuous real-valued function on* X, *a and b points of* X. *Then f takes on every value between $f(a)$ and $f(b)$.*

For, $f(X)$ is a connected subset of **R** (10.1.8), hence is an interval of **R** (10.1.9). This interval contains $f(a)$ and $f(b)$, hence all numbers in between.

10.2. Arcwise Connected Spaces

10.2.1. Definition. Let X be a topological space and $a, b \in X$. A continuous mapping f of $[0, 1]$ into X such that $f(0) = a$, $f(1) = b$ is called a *continuous path in* X with *origin a* and *extremity b*. If any two points of X are the origin and extremity of a continuous path, X is said to be *arcwise connected.*

10.2.2. Example. If E is a normed vector space, then E is arcwise connected. For, if $a, b \in E$, the mapping

$$t \mapsto f(t) = a + t(b - a)$$

of [0, 1] into E is continuous (8.1.8), and $f(0) = a$, $f(1) = b$. For example, $\mathbf{R}^n$ is arcwise connected.

▶ **10.2.3. Theorem.** *Let* X *be an arcwise connected topological space. Then* X *is connected.*

Let $x_0 \in X$. For every $x \in X$, let $f_x\colon [0, 1] \to X$ be a continuous path with origin x_0 and extremity x. Since [0, 1] is connected (10.1.9), the set $A_x = f_x([0, 1])$ is a connected subset of X (10.1.8). Now, $x \in A_x$, thus the union of the A_x is X. Since x_0 belongs to all of the A_x, X is connected (10.1.6).

10.3. Connected Components

10.3.1. Theorem. *Let* X *be a topological space, and* $x \in X$. *Among the connected subsets of* X *containing* x, *there exists one that is larger than all the others.*

There exists at least one such set, namely $\{x\}$. The union of all of the connected subsets of X containing x is connected (10.1.6) and is obviously the largest of the connected subsets of X containing x.

10.3.2. Definition. The subset of X defined in 10.3.1 is called the *connected component of* x in X.

10.3.3. Theorem. *Let* X *be a topological space.*

(i) *Every connected component of* X *is closed in* X.
(ii) *Two distinct connected components are disjoint. In other words, the different connected components of* X *form a partition of* X.

(i) Let A_x be the connected component of x. Then $\overline{A}_x$ is connected (10.1.7). But A_x is the largest connected subset of X containing x, therefore $\overline{A}_x = A_x$.

(ii) Let A_x, A_y be connected components that are not disjoint. Then $A_x \cup A_y$ is connected (10.1.6). Since $x \in A_x \cup A_y$, we have $A_x \cup A_y \subset A_x$, whence $A_y \subset A_x$. Similarly $A_x \subset A_y$, therefore $A_x = A_y$.

10.3.4. Examples. A connected space has only one connected component. The space $\mathbf{R} - \{0\}$ has two connected components, namely $(-\infty, 0)$ and $(0, +\infty)$.

Exercises

Chapter I

1. Let Δ be the set of $(x, y) \in \mathbf{R}^2$ such that $x^2 + y^2 \leq 1$. Let S be the set of points of $\mathbf{R}^2$ of the form $(x, 0)$ with $0 \leq x \leq 1$. Let $A = \Delta - S$. Find the interior, exterior, boundary and closure of A (relative to $\mathbf{R}^2$).

2. Let X be a topological space. If A is a subset of X, we denote by Bd(A) the boundary of A.

(a) Show that $\mathrm{Bd}(\mathring{A}) \subset \mathrm{Bd}(A)$, $\mathrm{Bd}(\bar{A}) \subset \mathrm{Bd}(A)$. Show by means of an example (try $X = \mathbf{R}$ and $A = \mathbf{Q} \cap [0, 1]$) that these three sets can be distinct.

(b) Let A, B be subsets of X. Show that

$$\mathrm{Bd}(A \cup B) \subset \mathrm{Bd}(A) \cup \mathrm{Bd}(B);$$

show by means of an example (try $X = \mathbf{R}$, $A = \mathbf{Q} \cap [0, 1]$ and $B = [0, 1] - A$) that these two sets can be distinct. If $\bar{A} \cap \bar{B} = \varnothing$, then $\mathrm{Bd}(A \cup B) = \mathrm{Bd}(A) \cup \mathrm{Bd}(B)$.

3. (a) Show that, on a set with two elements, there exist four topologies.

(b) On a finite set, the only separated topology is the discrete topology.

4. An open subset of **R** is the union of a sequence of pairwise disjoint open intervals.

5. One ordinarily identifies **R** with the subset $\mathbf{R} \times \{0\}$ of $\mathbf{R}^2$. Then $[0, 1]$ has nonempty interior relative to **R**, but empty interior relative to $\mathbf{R}^2$.

Chapter II

1. Let X be a topological space and A a nonempty subset of X. A subset V of X is called a neighborhood of A if there exists an open subset U of X such that $A \subset U \subset V$.

(a) The set of neighborhoods of A is a filter $\mathscr{F}$.

(b) Give a necessary and sufficient condition for the identity mapping of X into X to have a limit along $\mathscr{F}$ (assuming X is separated).

(c) Let $X = \mathbf{R}$, $A = \mathbf{N}$. Show that there does not exist a sequence $V_1, V_2, V_3, \ldots$ of elements of $\mathscr{F}$ such that every element of $\mathscr{F}$ contains one of the V_i.

2. Define a mapping f of $\mathbf{R}$ into $\mathbf{R}$ by

$$f(x) = (1 + e^x) \sin x.$$

Find the adherence values of f: (a) as $x \to -\infty$; (b) as $x \to +\infty$; (c) as $x \to 0$.

3. Let X be the set $\mathbf{R}$ equipped with the discrete topology. Show that the identity mapping of X into $\mathbf{R}$ is continuous, but is neither open nor closed.

4. Let X, Y be topological spaces, f a mapping of X into Y. The following conditions are equivalent: (a) f is continuous and closed; (b) $f(\overline{A}) = \overline{f(A)}$ for every subset A of X.

Chapter III

1. Let E be a topological space and $I = [0, 1]$. On the product space $E \times I$, consider the equivalence relation R whose classes are: (i) the sets with one element $\{(x, t)\}$, where $x \in E$, $t \in I$, $t \neq 1$; (ii) the set $E \times \{1\}$. The topological space $C = (E \times I)/R$ is called the cone constructed over E.

(a) For $x \in E$, denote by $f(x)$ the canonical image of $(x, 0)$ in C. Show that f is a homeomorphism of E onto $f(E)$.

(b) Show that E is separated if and only if C is separated.

2. Let X be a topological space, L a subset of X and $x \in L$. We say that L is locally closed at x if there exists a neighborhood V of x in X such that $L \cap V$ is a closed set in the subspace V. We say that L is locally closed in X if it is locally closed at each of its points.

(a) Show that the following conditions are equivalent: (i) L is locally closed; (ii) L is open in $\overline{L}$; (iii) L is the intersection of an open set and a closed set in X.

(b) The inverse image of a locally closed subset under a continuous mapping is locally closed.

(c) If L_1 and L_2 are locally closed in X, then $L_1 \cap L_2$ is locally closed in X.

(d) If L_1 is locally closed in L_2, and L_2 is locally closed in L_3, then L_1 is locally closed in L_3.

(e) Suppose that L is locally closed in X. Let $\mathcal{U}$ be the set of open subsets U of X such that $L \subset U$ and L is closed in U. Let F be the boundary of L with respect to $\overline{L}$. Then $X - F$ is the largest element of $\mathcal{U}$.

3. Let X, Y be topological spaces.

(a) Let $x, x_1, x_2, \ldots \in X$ and $y, y_1, y_2, \ldots \in Y$. If the sequence $((x_n, y_n))$ admits (x, y) as adherence value in $X \times Y$, then the sequence (x_n) (resp. (y_n)) admits x (resp. y) as adherence value in X (resp. Y).

(b) Show that there exists in $\mathbf{R}^2$ a sequence $((x_n, y_n))$ that admits no adherence value, even though each of the sequences (x_n) and (y_n) has an adherence value in $\mathbf{R}$.

4. The canonical projections of a product of topological spaces onto the factor spaces are open mappings.

5. Let X, Y be topological spaces, $A \subset X$ and $B \subset Y$. The following topologies on $A \times B$ coincide: (a) the topology induced by the product topology on $X \times Y$; (b) the product of the induced topologies on A and B.

6. On $\mathbf{R}^n$, define an equivalence relation $\mathcal{R}$ in the following way: $(x_1, x_2, \ldots, x_n)$ and $(y_1, y_2, \ldots, y_n)$ are equivalent if $x_i - y_i \in \mathbf{Z}$ for all i. Show that the quotient space $\mathbf{R}^n/\mathcal{R}$ is homeomorphic to $\mathbf{T}^n$.

7. Let p be the canonical mapping of $\mathbf{R}$ onto $\mathbf{T}$. Let f be the mapping $x \mapsto (p(x), p(x\sqrt{2}))$ of $\mathbf{R}$ into $\mathbf{T}^2$. Show that f is injective and continuous, but that f is not a homeomorphism of $\mathbf{R}$ onto $f(\mathbf{R})$.

8. Let X, Y be separated topological spaces, f a continuous mapping of X into Y. The graph of f is a closed subset of X × Y.

9. Let E be a topological space, F and G subsets of E such that G ⊂ F. For G to be closed in F, it is necessary and sufficient that $\overline{G} \cap F = G$, where $\overline{G}$ denotes the closure of G in E.

Chapter IV

1. In $\mathbf{R}^2$ equipped with the usual metric, let D be an open disc with center x_0 and radius $\alpha > 0$, and let A be a compact set contained in D. Show that there exists $\alpha' \in (0, \alpha)$ such that A is contained in the open disc with center x_0 and radius α'.

2. Let E be a separated space, $(x_1, x_2, \ldots)$ a sequence of points of E that tends to a point x of E. Show that $\{x, x_1, x_2, \ldots\}$ is compact.

3. Let E be a separated space. Suppose that for every set X, every filter base $\mathscr{B}$ on X, and every mapping f of X into E, f admits an adherence value along $\mathscr{B}$. Then E is compact.

4. The topological spaces (0, 1) and [0, 1] are not homeomorphic.

5. Let E_1, E_2 be nonempty topological spaces. If $E_1 \times E_2$ is compact, then E_1 and E_2 are compact.

6. Let $A = \mathbf{R}^{n+1} - \{0\}$. Define an equivalence relation $\mathscr{R}_n$ on A in the following way: two points x and y of A are equivalent if there exists $t \in \mathbf{R} - \{0\}$ such that $y = tx$. The quotient space $A/\mathscr{R}_n$ is denoted $\mathbf{P}_n(\mathbf{R})$ and is called the real projective space of dimension n.

(a) Let π be the canonical mapping of A onto $\mathbf{P}_n(\mathbf{R})$. Show that π is open.

(b) Show that π is not closed.

(c) Let Γ be the set of $(x, y) \in A \times A$ such that x is equivalent to y. Show that Γ is closed. From this, deduce that $\mathbf{P}_n(\mathbf{R})$ is separated.

(d) Let $\mathscr{S}_n$ be the equivalence relation on $\mathbf{S}_n$ obtained by restriction of $\mathscr{R}_n$. Show that the quotient space $\mathbf{S}_n/\mathscr{S}_n$ is compact.

(e) Let φ be the restriction of π to $\mathbf{S}_n$. Show that φ is continuous and defines a homeomorphism of $\mathbf{S}_n/\mathscr{S}_n$ onto $\mathbf{P}_n(\mathbf{R})$, so that $\mathbf{P}_n(\mathbf{R})$ is compact.

(f) Let g be the mapping of $\mathbf{S}_1$ into $\mathbf{R}_2$ such that $g(x, y) = (x^2 - y^2, 2xy)$ for $x, y \in \mathbf{R}$, $x^2 + y^2 = 1$. Show that $g(\mathbf{S}_1) = \mathbf{S}_1$ and that g defines a homeomorphism of $\mathbf{S}_1/\mathscr{S}_1$ onto $\mathbf{S}_1$, so that $\mathbf{P}_1(\mathbf{R})$ is homeomorphic with $\mathbf{S}_1$.

Chapter V

1. Show that the conclusion of 5.5.12 may fail if X is not assumed to be complete. (Take $X = \mathbf{Q}$. Let $r_1, r_2, \ldots$ be the elements of $\mathbf{Q}$ arranged in a sequence. Take $U_n = \mathbf{Q} - \{r_n\}$.)

2. Let X be a metric space and A, B, C subsets of X. Show that one does not necessarily have $d(A, C) \leq d(A, B) + d(B, C)$. (Take $X = \mathbf{R}$, $A = [0, 1]$, $B = [1, 2]$, $C = [2, 3]$.)

3. In $\mathbf{R}^n$, equipped with the usual Euclidean metric, the diameter of an open or closed ball is twice its radius.

4. Let X be a set. For $x, y \in X$, set $d(x, y) = 1$ if $x \neq y$, $d(x, y) = 0$ if $x = y$. Show that d is a metric on X and that the corresponding topology is the discrete topology. The diameter of a ball of radius <1 is 0.

5. Let X be a metric space, and $x, x_1, x_2, x_3, \ldots$ points of X. Show that the following conditions are equivalent: (i) $x_n \to x$; (ii) every subsequence of $(x_1, x_2, \ldots)$ has a subsequence tending to x.

6. In 5.5.10, there are four hypotheses: X complete, the F_n closed, the F_n decreasing, $\delta_n \to 0$. Show that if any one of these hypotheses is omitted, it can happen that $F_1 \cap F_2 \cap \cdots$ is empty.

7. Show that in 5.5.13, f may fail to exist if f' is only assumed to be continuous (but not uniformly continuous). (Take $X = Y = \mathbf{R}$, $X' = \mathbf{R} - \{0\}$, $f(x) = \sin(1/x)$ for $x \in X'$.)

8. Let X be a metric space, A a closed subset of X, B a compact subset of X. Assume that $A \cap B = \varnothing$. Show that $d(A, B) > 0$.

9. Let $f_1, f_2, \ldots$ be continuous functions ≥ 0 on [0, 1], such that: (α) $f_1 \geq f_2 \geq f_3 \geq \cdots$; ($\beta$) the only continuous function ≥ 0 on [0, 1] that is majorized by all of the f_n is the function 0.

(a) For every $x \in [0, 1]$, let $l(x) = \lim_{n\to\infty} f_n(x)$. For every integer $n \geq 1$, let $O_n = \{x \in [0, 1] \mid l(x) < 1/n\}$. Show that O_n is a dense open set in [0, 1].

(b) Show that l vanishes on a dense subset of [0, 1].

Chapter VI

1. Let C be the set of continuous real-valued functions on [0, 1], equipped with the metric of uniform convergence. If n is an integer > 0, we denote by A_n the set of $f \in C$ for which there exists $t \in [0, 1 - 1/n]$ with the following property:

$$(*) \qquad |f(t') - f(t)| \leq n(t' - t) \quad \text{for all} \quad t' \in [t, t + 1/n].$$

(a) Let $f \in C$ be differentiable at at least one point of [0, 1). Show that $f \in A_n$ for some n.

(b) Let $(f_1, f_2, \ldots)$ be a sequence of elements of A_n tending uniformly to an element f of C. Assume that there exists in $[0, 1 - 1/n]$ a sequence $t_1, t_2, \ldots$ having a limit t, such that

$$|f_i(t') - f_i(t_i)| \leq n(t' - t_i) \quad \text{for all} \quad t' \in [t_i, t_i + 1/n].$$

(α) Show that $f_i(t_i) \to f(t)$.

(β) Show that $|f(t') - f(t)| \leq n(t' - t)$ for $t' \in (t, t + 1/n)$.

(γ) Show that t has the property (*) relative to f.

(c) Show that A_n is closed in C.

(d) Let $f \in C$ be a continuously differentiable function and let $\varepsilon > 0$. Let n be an integer > 0. Set $M = \sup_{0 \leq x \leq 1} |f'(x)|$.

Let $0 = \alpha_0 < \alpha_1 < \alpha_2 < \cdots < \alpha_{p-1} < \alpha_p = 1$ be real numbers such that each interval $[\alpha_i, \alpha_{i+1}]$ has length $\leq \varepsilon/(M + 2n)$. Let g be the function that is linear on each of the intervals $[\alpha_i, \alpha_{i+1}]$, and is such that

$$g(\alpha_0) = g(\alpha_2) = g(\alpha_4) = \cdots = 0,$$
$$g(\alpha_1) = g(\alpha_3) = g(\alpha_5) = \cdots = \varepsilon.$$

(α) Show that $|g(t') - g(t)| \geq (M + 2n)|t' - t|$ if t and t' belong to the same interval $[\alpha_i, \alpha_{i+1}]$.
(β) Show that $d(f, f + g) \leq \varepsilon$.
(γ) Show that $f + g \notin A_n$. (Argue by contradiction.)

(e) Show that the interior of A_n (relative to C) is empty. (Argue by contradiction. One uses (d) and the fact that every continuous real-valued function on [0, 1] is the uniform limit of continuously differentiable functions; on this subject, see 7.5.5.)

(f) Show that the intersection of the complements $C - A_1, C - A_2, \ldots$ is non-empty. Show that $A_1 \cup A_2 \cup \cdots \neq C$.

(g) Show that there exists an $f \in C$ that is not differentiable at any point of [0, 1].

2. This exercise is a commentary on Dini's theorem.

(a) Construct on [0, 1] an increasing sequence of continuous real-valued functions $f_1, f_2, \ldots$ that tends simply to a function f that is not continuous (so that f_n does not tend uniformly to f).

(b) Construct on [0, 1] a sequence of continuous real-valued functions that tends simply to 0 but does not tend uniformly to 0.

(c) Construct on **R** an increasing sequence of continuous real-valued functions that tends simply to the function 1 but does not tend uniformly to 1.

3. Let F be a continuous real-valued function on $[0, 1]^3$. Let $C = \mathscr{C}([0, 1], [0, 1])$. Throughout the exercise, one utilizes the metric of uniform convergence.

(a) Let $f \in C$. Show that for every $s \in [0, 1]$, the function $t \mapsto F(s, t, f(t))$ on [0, 1] is continuous. Set

$$g(s) = \int_0^1 F(s, t, f(t))\, dt.$$

Show that $g \in \mathscr{C}([0, 1], \mathbf{R})$.

(b) As f varies, we have thus defined a mapping $f \mapsto g$ of C into $\mathscr{C}([0, 1], \mathbf{R})$. Let φ denote this mapping. Show that φ is uniformly continuous.

(c) Show that $\varphi(C)$ has compact closure in $\mathscr{C}([0, 1], \mathbf{R})$.

Chapter VII

1. Let $f: \mathbf{R} \to [0, +\infty)$ be a lower semicontinuous function. Show that f is the upper envelope of a family of continuous functions ≥ 0.

2. Let X be a topological space, $x_0 \in X$, and f, g finite numerical functions on X that are lower semicontinuous at x_0. Suppose that $f + g$ is continuous at x_0. Show that f and g are continuous at x_0.

3. Let $(u_1, u_2, u_3, \ldots)$ be a sequence of real numbers such that $u_{m+n} \leq u_m + u_n$ for $m, n \geq 1$. Let

$$l = \limsup_{n\to\infty} \frac{u_n}{n}, \qquad l' = \liminf_{n\to\infty} \frac{u_n}{n}.$$

(a) Show that $u_{i_1+i_2+\cdots+i_n} \leq u_{i_1} + u_{i_2} + \cdots + u_{i_n}$ for all $i_1, \ldots, i_n$, and that $u_{mn} \leq mu_n$ for all positive integers m, n.

(b) Fix an integer $p \geq 1$. Let $\alpha = \sup(0, u_1, u_2, \ldots, u_{p-1})$.

(α) For every integer $n \geq 1$, write $n = k_n p + r_n$ with $0 \leq r_n < p$ (Euclidean division of n by p). Show that

$$\frac{u_n}{n} \leq \frac{k_n p}{n} \cdot \frac{u_p}{p} + \frac{\alpha}{n}.$$

(β) Show that $l \leq u_p/p$. What can be deduced about l' from this?

(c) Show that u_n/n has a limit λ as $n \to \infty$, and that $u_n/n \geq \lambda$ for all n.

4. Consider the topological space

$$X = [0, 1] \times [0, 1] \times [0, 1] \times \cdots.$$

An element of X is an infinite sequence $(x_1, x_2, x_3, \ldots)$ of numbers in $[0, 1]$.

(1) Let n be an integer ≥ 1. If $f \in \mathscr{C}([0, 1]^n, \mathbf{R})$, one defines a function g on X by setting

$$g(x_1, x_2, x_3, \ldots) = f(x_1, x_2, \ldots, x_n)$$

for any $(x_1, x_2, x_3, \ldots) \in X$. Show that $g \in \mathscr{C}(X, \mathbf{R})$. Let C_n be the set of continuous functions on X obtained in this way as f runs over $\mathscr{C}([0, 1]^n, \mathbf{R})$.

(2) Show that $C_1 \subset C_2 \subset C_3 \subset \cdots$.

(3) Show that $C_1 \cup C_2 \cup C_3 \cup \cdots$ is dense in $\mathscr{C}(X, \mathbf{R})$ for the topology of uniform convergence.

5. For $a, b, c \in \mathbf{R}$ and $a < 0$, denote by $f_{a,b,c}$ the function $x \mapsto e^{ax^2+bx+c}$ on $\mathbf{R}$. Show that every continuous complex-valued function on $\mathbf{R}$ tending to 0 at infinity is the uniform limit on $\mathbf{R}$ of linear combinations of the functions $f_{a,b,c}$.

Chapter VIII

1. Let $(e_1, e_2, \ldots)$ be the canonical orthonormal basis of l^2. Show that this sequence has no convergent subsequence. From this, deduce that l^2 is not locally compact.

2. Show that there exists no scalar product on $l^1_{\mathbf{C}}$ for which the corresponding norm is the norm of 8.1.5. (Show that the parallelogram law fails.) Analogous question for $l^\infty_{\mathbf{C}}$.

3. Let E be a separated pre-Hilbert space, C a complete convex subset of E, x a point of E. Show that there exists one and only one $y \in C$ such that $d(x, y) = d(x, C)$.

4. For every $x = (x_1, x_2, \ldots) \in l^\infty_{\mathbf{R}}$, define a linear form f_x on $l^1_{\mathbf{R}}$ by the formula

$$f_x((y_1, y_2, y_3, \ldots)) = x_1y_1 + x_2y_2 + x_3y_3 + \cdots$$

for all $(y_1, y_2, y_3, \ldots) \in l^1_{\mathbf{R}}$ (note that the series on the right side is absolutely convergent). Show that $f_x \in (l^1_{\mathbf{R}})'$ and that the mapping $x \mapsto f_x$ of $l^\infty_{\mathbf{R}}$ into $(l^1_{\mathbf{R}})'$ is a linear bijection such that $\|f_x\| = \|x\|$ for all $x \in l^\infty_{\mathbf{R}}$.

5. Let E be a normed space and $u \in \mathscr{L}(E)$. Show that $\|u^n\|^{1/n}$ has a limit as $n \to \infty$. (Use Exercise 3 of Chapter VII.)

6. Let $f_1, f_2, \ldots$ be continuous complex-valued functions on [0, 1]. Consider the following conditions: (a) $f_n \to 0$ uniformly; (b) $f_n \to 0$ in mean square; (c) f_n tends to 0 simply.

Then (a) $\Rightarrow$ (b), (a) $\Rightarrow$ (c). Show that the implications (b) $\Rightarrow$ (a), (c) $\Rightarrow$ (b), (b) $\Rightarrow$ (c) are false.

Chapter IX

1. If, in a normed space E, every absolutely convergent series is convergent, then E is complete.

2. Let E be a Banach space and $u \in \mathscr{L}(E)$. Show that the series

$$1 + u + \frac{u^2}{2!} + \frac{u^3}{3!} + \cdots$$

is absolutely convergent in $\mathscr{L}(E)$. Its sum is denoted e^u. Show that e^u is bicontinuous, with inverse e^{-u}. Show that $e^{u+v} = e^u e^v$ if $u, v \in \mathscr{L}(E)$ and $uv = vu$. Show that $\|e^u\| \le e^{\|u\|}$.

3. Let $a, b \in [0, 1)$. Show that the family $(a^m b^n)_{(m,n) \in \mathbf{N} \times \mathbf{N}}$ is summable in **R**.

4. Let $(a_i)_{i \in I}$ be a summable family of numbers ≥ 0. Then the family $(a_i^2)_{i \in I}$ is summable.

5. Let $(x_1, x_2, x_3, \ldots)$ be a sequence of real numbers. Suppose that for every sequence $(y_1, y_2, y_3, \ldots)$ of real numbers tending to 0, the series $x_1 y_1 + x_2 y_2 + x_3 y_3 + \cdots$ is convergent. Show that $|x_1| + |x_2| + |x_3| + \cdots < +\infty$.

Chapter X

1. Let X, Y be topological spaces. Consider the following conditions: (a) X and Y are connected; (b) X $\times$ Y is connected.

Then (a) $\Rightarrow$ (b). If X and Y are nonempty, (b) $\Rightarrow$ (a).

2. Let Ω be an open subset of $\mathbf{R}^n$. The following conditions are equivalent: (a) Ω is connected; (b) Ω is arcwise connected.

3. Let X be a topological space, and $a, b \in$ X. The relation:

there exists a continuous path in X with origin a and extremity b

is an equivalence relation in X. The equivalence classes for this relation are called the arcwise-connected components of X.

4. A topological space X is said to be locally connected if every point admits a fundamental system of connected neighborhoods. If this is the case, then all of the connected components of X are open in X.

5. The topological spaces X = (0, 1) and Y = [0, 1) are not homeomorphic (compare the complements of a point in X and in Y).

6. One wants to show that it is impossible to find a sequence of pairwise disjoint, nonempty closed subsets $F_1, F_2, \ldots$ of $[0, 1]$ whose union is $[0, 1]$. One argues by contradiction, on supposing that such subsets have been constructed.

(a) If $\mathrm{Bd}(F_n)$ denotes the boundary of F_n in $[0, 1]$, show that $F = \mathrm{Bd}(F_1) \cup \mathrm{Bd}(F_2) \cup \cdots$ is closed in $[0, 1]$.

(b) Show that the interior of $\mathrm{Bd}(F_n)$ in F is empty. Deduce a contradiction from this by applying Baire's theorem.

Index of Notations

(The references are to paragraph numbers)

Index of Terminology

(The references are to paragraph numbers)

Undergraduate Texts in Mathematics

continued from ii

Martin: The Foundations of Geometry and the Non-Euclidean Plane.
1975. xvi, 509 pages. 263 illus.

Martin: Transformation Geometry: An Introduction to Symmetry.
1982. xii, 237 pages. 209 illus.

Millman/Parker: Geometry: A Metric Approach with Models.
1981. viii, 355 pages. 259 illus.

Owen: A First Course in the Mathematical Foundations of Thermodynamics
1984. xvii, 178 pages. 52 illus.

Prenowitz/Jantosciak: Join Geometrics: A Theory of Convex Set and Linear Geometry.
1979. xxii, 534 pages. 404 illus.

Priestly: Calculus: An Historical Approach.
1979, xvii, 448 pages. 335 illus.

Protter/Morrey: A First Course in Real Analysis.
1977. xii, 507 pages. 135 illus.

Ross: Elementary Analysis: The Theory of Calculus.
1980. viii, 264 pages. 34 illus.

Sigler: Algebra.
1976. xii, 419 pages. 27 illus.

Simmonds: A Brief on Tensor Analysis.
1982. xi, 92 pages. 28 illus.

Singer/Thorpe: Lecture Notes on Elementary Topology and Geometry.
1976. viii, 232 pages. 109 illus.

Smith: Linear Algebra.
1978. vii, 280 pages. 21 illus.

Smith: Primer of Modern Analysis
1983. xiii, 442 pages. 45 illus.

Thorpe: Elementary Topics in Differential Geometry.
1979. xvii, 253 pages. 126 illus.

Troutman: Variational Calculus with Elementary Convexity.
1983. xiv, 364 pages. 73 illus.

Whyburn/Duda: Dynamic Topology.
1979. xiv, 338 pages. 20 illus.

Wilson: Much Ado About Calculus: A Modern Treatment with Applications Prepared for Use with the Computer.
1979. xvii, 788 pages. 145 illus.